The Carbon War

GLOBAL WARMING AND THE END OF THE OIL ERA

The Carbon War

GLOBAL WARMING AND THE END OF THE OIL ERA

JEREMY LEGGETT

ROUTLEDGE
NEW YORK

To Jess

Published in 2001 by
Routledge
29 West 35th Street
New York, NY 10001

First published in 1999 by Penguin Books Ltd.,
and reprinted by arrangement with Penguin Books Ltd.
First Routledge hardback edition, 2001
First Routledge paperback edition, 2001

Routledge is an imprint of the Taylor & Francis Group.

Printed in the United States of America on acid-free paper.

Library of Congress Cataloging-in-Publication data

Leggett, Jeremy K.
The carbon war : global warming and the end of the oil era / Jeremy Leggett.
p. cm.
Includes bibliographical references and index.
ISBN 0-415-93101-0 — ISBN 0-415-93102-9 (pbk.)
1. Global warming—Government policy. 2. Fossil fuels—Environmental aspects. I. Title.

QC981.8.C5 L45 2001
363.738'742—dc21 2001018158

10 9 8 7 6 5 4 3 2 1

7/01

Contents

Prologue

The Royal School of Mines at the Imperial College of Science, Technology and Medicine is an elite training house for oil and mining companies. There, within its Victorian corridors in Kensington, for more than a decade, I taught the ways and culture of the hunt for oil. I helped to turn out petroleum geologists and petroleum engineers in their hundreds. Along the way, I supplemented my income by consulting for the companies whose future servants I was helping to train. It seems scarcely credible to me now, but once I was able to buy my daughter a horse. Two, if I am honest. One summer I actually worked for an oil company. I was a suit-and-tie commuter in Tokyo, the sole foreigner in a Kasumigaseki skyscraper owned by Japex, the Japan Petroleum Exploration Company. The only thing I was excused from was daily performance of the company exercises.

In truth, I had discovered a great romance. Looking back, I have a fancy now that it stemmed from something primeval. I remember the hunter's thrill I felt in Baluchistan, watching smears of oil seeping from the ground. I felt the same thing in that Tokyo office, looking at a possible reservoir on a seismic record or a satellite image. Then there were the hunter's weapons. The ships and trucks pumping seismic energy into the ground, building pictures of the hidden subsurface. The drill rigs and down-hole instrument packages probing for the quarry. On top of that were the pipelines and the supertankers carrying the object of the hunt to market – via oil refineries, those most complex of Meccano constructs – where finally, of course, the prize could be burned: in engines, all kinds of fascinating engines.

At the time I loved it all. I was a young academic in an industry-focused university, surrounded by oil-industry megastars. The professor of petroleum geology at the Royal School of Mines was a retired Chief Geologist from BP, a legend in the business. When he first arrived I would talk to him in the common room, quite faint with admiration. His predecessor, before my time, was another legend. He had found an oilfield in Pakistan, and taken a financial stake in the small company that drilled it. When the well had hit oil, he threw an impromptu champagne party for half the faculty in a luxury Kensington hotel. Lectures were cancelled for days afterwards.

Although my teaching and consultancy had much to do with oil, my research was in what academics tend to refer to as blue skies: knowledge which has no obvious immediate commercial relevance. My subject was the geological history of oceans. How I loved that. In the modern oceans, I studied the unfolding history of the planet from sediment cores drilled under as much as five kilometres of water, recording in their thin layers stories of climate and life on a dynamic Earth over hundreds of millions of years. I studied the deposits of ancient oceans, lifted by the forces of plate tectonics from their deep-water origins to their present-day locations in mountains that mark where continents once collided. My work took me on research ships in the Pacific and to the mountain belts of Europe, Japan and Pakistan. I worked in particular with Japanese and French scientists, learning to appreciate such novelties as sashimi and Bordeaux. I was one happy camper.

But during the mid-1980s I began to notice a series of worrying papers and articles appearing in the scientific journals about the build-up of greenhouse gases in the atmosphere. Translating the carefully coded language I was reading, it seemed that atmospheric physicists were becoming concerned that burning so much coal, oil and gas – the carbon, or fossil, fuels – risked turning up the planetary thermostat too high, destablizing global climate. I knew a thing or two about climate, from the bottom up, as it were, having studied so many stories of the past in the oceanic record. I knew how slowly the natural climatic rhythms of the planet worked. Now the people looking at climate from the top down were saying there was a danger of changes of a rapidity never before seen. I felt my own concern rising by the month.

1988 was the year that broke the mould. The news about global warming that summer became impossible to ignore. I felt my sense of mission, future and professional identity eroding with every new report I read. At first, I tried to reconcile the information with what I saw as my life's calling. I petitioned my colleagues in the Royal School of Mines to let me start giving a few environmental science lectures. The students, I argued, should be aware of what was brewing in the atmospheric sciences. Quite apart from the question of whether or not it might be a good idea to give them an all-round education, this environmentalism business might hold implications for their job prospects. My colleagues reluctantly agreed, but immediately took to referring to the lectures as 'Dr Leggett's liberal-studies classes'.

The following year, the fault lines growing by increments in my sense of professional identity reached failure point. One day early in 1989, I stood in front of a class of forty undergraduates giving a lecture on an oilfield in California. I had an interesting hunter's tale to tell that morning: how the oil had been trapped below ground over millions of years; how Chevron had discovered the monster oilfield long after many companies had concluded there was nothing there; the technical tricks they had used; the industrial espionage they had had to evade from their sister companies in order to keep their discovery secret until they had bought up as much as they could of the rest of the oilfield. The rows of young people sat listening quietly. And as I stared down at the upturned faces, I suddenly had the feeling that I could not go on. That day, the tension between my growing environmental concerns and my job description in the Royal School of Mines came to a head. I went straight back to my office and turned to the job pages in *New Scientist* magazine.

The events of 1988 had burrowed further into my conscience than I had admitted to myself. In June, a group of climate-change and ozone-depletion experts meeting in Toronto had issued a statement in which they concluded that the global warming being primed by the burning of oil, coal and gas could combine with acid rain and loss of ozone to unleash consequences, as they put it, second only to those of nuclear war. It seemed entirely logical. In the aftermath of the shock discovery of the Antarctic ozone hole in 1985, the world knew by then that CFC gases destroyed ozone in the atmosphere, exposing life on the planet to growing amounts of ultraviolet radiation. We also knew that greenhouse gases trapped heat. There was no scientific doubt about that, then as now: in fact, if it were not for a natural greenhouse effect, the world would be too cold to support life. And we knew, because measurements showed it the world over, that atmospheric concentrations of greenhouse gases were rising rapidly, so enhancing the natural greenhouse effect. Carbon fuel combustion was undoubtedly the main culprit: it produced most of the main greenhouse gas, carbon dioxide. The decade of the 1980s had already been significantly warmer, on the global average, than any other in more than a hundred years of records. Climate and the global average temperature were naturally variable, to be sure, but suspicions were growing among the scientific community in 1988 that the hot years of the 1980s were the first faint footprint of global warming.

The northern-hemisphere summer of 1988 was rife with potential foretastes of what lay ahead if we kept turning up that thermostat. In the US Midwest an appalling drought spread misery through the farmlands. In Yellowstone Park, with drought at its worst for over a hundred years, the worst-ever forest fires saw 25,000 firefighters battling walls of flame up to 60 metres high. Leading NASA scientist Jim Hansen testified in Congress that it was time to stop waffling and say that global warming had already started. UK Prime Minister Margaret Thatcher, giving a speech to the Royal Society in London, summarized precisely what I had begun to fear. 'We may have unwittingly begun a massive experiment with the system of the planet itself,' she warned. Here was a woman not otherwise known for eco-doom-mongering.

And there I stood, that day, giving my lecture on the giant offshore California oilfield, once again teaching the students new tricks in the search for oil, as though concerns about the global environment – my occasional liberal-studies classes notwithstanding – were somehow just a sideshow. Until then, my entire professional life had been dedicated to training young people like these to go forth and find fossil fuels, to add carbon as heat-trapping carbon dioxide and methane to the atmosphere. To quite literally fuel a threat to the future, and risk bequeathing their children an uninhabitable world. It had to stop.

Within weeks of scanning *New Scientist*'s jobs pages I found that the international environmental group Greenpeace wanted a scientist to work in their UK office to give technical advice to their campaigners. The issues were becoming increasingly complex, they said. The penalties for technical errors were becoming increasingly severe for environmental groups. Credibility was all. So it was that Greenpeace offered me the chance of moving from one of the most conservative universities in the world to one of the most radical environmental groups.

I jumped at it.

Acknowledgements

First, to my parents, daughter and sister: Dennis, Audrey, Jess and Jenny. I regard my vocational life before the carbon war, with the benefit of hindsight, as a form of extended boot camp for life in the trenches. They were so supportive and loving through both phases. They kept me rooted, tolerated my frequent air miles, and generally made my efforts possible in so many ways. Second, to my uncle, Dick, who essentially bankrolled my years as an environmentalist, was never less than absorbed with progress, and commented helpfully on early drafts of this book. Third, to all my former colleagues in Greenpeace, the other environment groups, and our open and covert supporters in government and industry. There were so many of them who contributed to this story: many more than I have been able to name in these pages. Fourth, to Sir Crispin Tickell for being so inspirationally wise about it all, and making so much effort to try to pass this on to me and so many other people. Fifth, to Ross Gelbspan for editorial assistance and for deploying his skills as an investigative journalist where too few of his fellows saw fit to tread. Sixth, to Rolf Gerling and Stephan Schmidheiny, for the continuing support – financial and advisory – that is now allowing me to have a shot at trying to help kick-start the solar revolution. Seventh, to my colleagues at Oxford University's Environmental Change Unit, in particular for putting up with so many absences. Eighth, to Aki Maruyama, for a friendship which transcends the bickering of nations. And finally, to my editor, Caroline Pretty. To the extent that this book is readable, she is mostly responsible. Nobody but me, however, bears responsibility for the content.

Abbreviations

AGBM	Ad-hoc Group on the Berlin Mandate (comprising all governments negotiating the Kyoto Protocol)
AOSIS	Alliance of Small Island States
BCSEF	Business Council for a Sustainable Energy Future
COP	Conference of Parties (of the Framework Convention on Climate Change)
GCC	Global Climate Coalition
G77	The Group of Seventy-seven (an umbrella group of developing nations at the UN)
HCFCs	hydrochlorofluorocarbons (a group of potent greenhouse gases)
HFCs	hydrofluorocarbons (a group of potent greenhouse gases)
ICC	International Chamber of Commerce
IPCC	Intergovernmental Panel on Climate Change
JUSCANZ	Japan, the USA, Canada, Australia and New Zealand (an informal group of countries resistant to the European Union's efforts to achieve early emissions reductions during the climate negotiations)
NGO	Non-Governmental Organization (such as business lobby groups and environment groups)
OECD	Organization for Economic Cooperation and Development
OPEC	Organization of Petroleum-Exporting Countries
UNEP	United Nations Environment Programme
WBCSD	World Business Council for Sustainable Development

I

The Early Warning

October 1989–December 1990

OCTOBER 1989, BERLIN

My first sight of the Berlin Wall etched itself on my mind for life. The killing grounds looked surreal, neon-lit in a misty, 1 a.m. light as I rattled past on an empty train. Through bleary eyes I stared at the concrete ramparts either side of the sandy no man's land, the drapes of coiled razor wire, the machine-gun towers trained on that murderous hundred-metre gap between tyranny and hope. It was my first time in Berlin, and I felt an acute sense of unreality. My world, and the world in general, were both changing with bewildering speed. I was about to fly to glasnost-gripped Moscow to work with Russian colleagues in a newly formed Greenpeace office, the setting-up of which Mikhail Gorbachev had himself sanctioned. The Cold War, with its threat of nuclear annihilation, seemed to be miraculously evaporating. More and more people were realizing – Gorbachev apparently among them – that we were entering a new era of security threat. The Soviet leader had been talking about the dangers of global warming in his speeches for several years.

Looking at the sombre course of the Wall east of the Friedrichstrasse in October 1989, my abiding thought was how anachronistic it was. But not for a moment did I think that within a matter of weeks people would be taking sledgehammers to those machine-gun towers. Before the Berlin Wall came down, anyone predicting its fall would have been laughed at. Even when the Wall had fallen, no pundit that I am aware of came close to predicting what subsequently happened once that particular engine of change had been kick-started. Similarly, if you predict today that the world is on track for a collapse in the use of carbon fuels, that huge amounts are going to be staying in the ground unburned – that mushrooming multibillion dollar markets in solar energy will emerge before the next century is very old at all – then you can expect to be greeted with the mirth of experts.

But so it was with the pundits, in October 1989, about the toppling of the Berlin Wall and the fall of Soviet communism. Neither institution had been quickly undermined. They had shown few signs that their foundations were crumbling. But fall they did, and when they went they went quickly.

MAY 1990, BERKSHIRE, UK

The first crack in the foundations of the carbon era is traceable to events in a country hotel in Berkshire during the spring of 1990. In this rural retreat, a hundred scientists gathered for three days to put the finishing touches to a document destined to become one of the most important scientific reports ever compiled. They came from government and university labs all over the world. They picked their location hoping for a seclusion befitting their sober task. But long before they had completed their deliberations, TV news crews and radio and print journalists were thronging in the lobby of the hotel, waiting impatiently for the scientists to emerge from the conference hall where they were at work. From a seat at the back of the hall, I knew I was watching history in the making.

In 1988, faced with growing concerns about climate change, the United Nations General Assembly had set up a panel to advise governments on the issue. The Intergovernmental Panel on Climate Change, or the IPCC as it was to become universally known, represented a consultation process unprecedented both in size and scope. Its mission was to pool the opinions of as many scientists and policy experts as possible, in as many countries as possible, and to thrash out over the next 18 months consensus reports on the science of global warming, the probable impacts, and the potential policy responses.

I was not alone in observing the final drafting of the IPCC's historic first Scientific Assessment Report. In other seats at the back of the room sat eleven scientists from the oil, coal and chemical industries, including two from Exxon, one from Shell and one from BP. Although they were allowed to take part as observers, this role was loosely defined, since they were permitted to make suggestions for wording as the text evolved. So too were I and the one other suitably qualified environment-group scientist present that day, Dr Dan Lashof of the Washington-based Natural Resources Defense Council.

The scientists were now working on the most important few paragraphs they would produce – the summary. Dr Brian Flannery, representing the International Petroleum Industries' Environmental Conservation Association, but on the payroll of Exxon, took the microphone. The draft, he reminded the room, said that 60 to 80 per cent cuts would be needed in carbon dioxide emissions in order to stabilize atmospheric concentrations of the gas. But this, he felt, required clarification in the light of all the uncertainties about the behaviour of carbon in the climate system.

Scientists from the UK, Germany and the USA – some of the most eminent climatologists in the world – now spoke. Nobody agreed with Flannery. Of course there were uncertainties, but, if your goal was to stabilize atmospheric concentrations of carbon dioxide, those uncertainties did not undermine the need for deep cuts in emissions.

Flannery took the microphone again. 'The range of model results isn't any better justified now than it was ten years ago,' he asserted, a didactic edge appearing in his voice. 'The range is quite scientifically uncertain. This should be stated as such in the executive summary.'

A leading climate modeller at the UK Met Office, frowning, waved his arm to attract the chairman's attention. 'I'd like to dispute that, Mr Chairman. The range is much better than it was three years ago, much less ten.' The Met Office man looked annoyed.

Others agreed. The discussion moved on.

The man with the most difficult and most crucial job that day was the chairman, Dr John Houghton, director-general of the Met Office. Houghton came to a critical sentence in the executive summary. 'Can we say we are *certain* that greenhouse-gas emissions at present rates will lead to warming?' he asked.

He was answered by a roomful of nodding heads.

*

The next day, the UK Prime Minister was due to hold a press conference at the Met Office, not far from the scientists' retreat. John Houghton had left the hotel the previous evening to brief her on the content of the scientists' report.

Margaret Thatcher was not a woman known for her concern about matters ecological. But today things were to change. Adopting one of her most dramatic expressions, the prime minister proceeded to rewrite a key paragraph of her place in history. 'Today,' she told the scribbling British press corps, 'with the publication of the report of the Intergovernmental Panel on Climate Change, we have an authoritative early-warning system: an agreed assessment from some three hundred of the world's leading scientists of what is happening to the world's climate. They confirm that greenhouse gases are increasing substantially as a result of man's activities, that this will warm the Earth's surface with serious consequences for us all.' It was, she said, a report of historic significance. What it predicted would affect our everyday lives.

She moved on to the impacts. 'There would surely be a great migration of population away from areas of the world liable to flooding, and from areas of declining rainfall and therefore of spreading desert. Those people will be crying out not for oil wells but for water.'

The next day, looking at the banner headlines in the morning papers, you would have been forgiven for thinking that Martians had invaded the planet. 'RACE TO SAVE OUR WORLD,' shrieked the *Daily Express*, the government's favourite tabloid. 'Britain takes lead in crusade against greenhouse effect,' the subhead announced proudly. All the other tabloids ran headlines in the same vein. Although the weightier papers did not adopt quite the same apocalyptic tone, they came close.

The authoritative early warning, bad though it was, could have been worse. The complexity of the climate system is such that there are many scientific uncertainties about the enhancement of the natural greenhouse effect. This makes the issue, at heart, one of threat assessment. The world may just get lucky, and find the resulting climate changes not quite as bad as most scientists estimate. Or the dice may roll unkindly: the changes could be a whole lot worse.

Dan Lashof and I went into that crucial 1990 IPCC meeting with great hopes that the worst-case analysis would be spelt out more starkly than it had been in the review copy of the report sent to attendees ahead of the meeting. We submitted papers to the IPCC, itemizing our concerns. What worried us most were the feedbacks in the climate system – the processes that can be triggered in a warming world which either amplify the warming (positive feedbacks) or suppress it (negative feedbacks). Our concern was that the former might end up swamping the latter.

We were far from alone in that fear. Such concerns had been fairly well explored in the scientific journals by this time. For example, a warming world could trigger extra emissions of greenhouse gases from the vast repositories of carbon in nature: from warming oceans, drying soils, dying forests, melting permafrost. These feedbacks were difficult – indeed often impossible – to quantify, and were for the most part excluded from computer models of climate. But, we wondered, shouldn't their role in the climate threat assessment be more clearly flagged? Shouldn't the worst-case analysis – a synergistic dominance of such feedbacks, uncompensated by negative feedbacks – be clearly spelt out for policymakers?

The section on confidence in predictions in the 1990 report included a carefully worded conclusion about natural sources of greenhouse gases, such as rotting vegetation, and 'sinks' where carbon dioxide is absorbed from the atmosphere, such as growing forests and plankton in the ocean. Both sources and sinks were sensitive to change in climate, the report read, so they might substantially modify future concentrations of greenhouse gases in the atmosphere. 'It appears likely,' the summary read, 'that, as climate warms, these feedbacks will lead to an overall increase, rather than decrease, in natural greenhouse gas abundances. For this reason, climate change is likely to be greater than the estimates we have given.'

But that was all the draft report said. The policymakers were left to read between the lines. They would have a difficult job. The 3°C rise in global average temperature estimated in the mid-range of the climate-model forecasts for the next century was described throughout the report as the 'best-guess' estimate.

Just as Exxon's Brian Flannery and other industry scientists tried throughout that meeting to water down the IPCC science assessment, so I and Dan Lashof tried to beef it up with references to the potential for feedback amplifications of warming. Choosing my moment as best I could in the final session, I urged the scientists to mention specifically what in principle the very worst case might be for a world where emissions were not cut deeply – a runaway, unstoppable greenhouse effect. Policymakers should have this spelt out for them clearly, I argued. If they understood the worst case, they would be more likely to buy insurance against it in the policies they came up with.

John Houghton asked the meeting, with clear reluctance, if there was any support for this view. An Austrian climatologist volunteered that in his opinion there was no way that a runaway greenhouse effect was possible. In any event, said a sea-level expert from the University of East Anglia, the media would seize on it and sensationalize it. That was all Houghton needed, and he dealt with me as circumspectly as he had earlier dealt with the oil and coal industry participants. Exxon's Brian Flannery fired the parting shot. This was simply scaremongering, he said.

Later, we came to a section in the report where specific biological feedbacks were described. The Director of Marine Sciences at the UK Government's Natural Environment Research Council wanted to strengthen a reference to one potential feedback, known as the plankton-multiplier effect. North Atlantic ocean circulation might turn off as a result of global warming, he said. If that happened, the implications for reduced phytoplankton productivity, and consequent boosting of atmospheric carbon dioxide, would be serious. He wanted that made clear.

I took the plunge again. That was the kind of thing that made it so important to spell out the potential sum of the prospective positive feedbacks, I said. There were a number of feedbacks like the one the Research Council man had stressed. For example, some reference should be made to the potential size, and vulnerability, of the methane hydrate reservoir in the Arctic.

Methane hydrates are ice-like substances which form under the permafrost and in the Arctic Ocean, and lock up a vast amount of methane under pressure. Warm them up, and they would be adding significantly to all the methane being emitted by humankind from gas pipelines, rice paddies and other places.

I could see frowns aplenty round the room. It was now the turn of Bob Watson of NASA, the leading US scientist at the meeting, to guard the 'best guess'. Watson had headed the team which a few years earlier had proved the link between CFCs and ozone depletion in the Antarctic ozone hole. 'I have a problem with this,' he said. 'We mustn't give policymakers the impression that there's no point. We don't win that way.'

Houghton seized on his point. 'Yes,' he said. 'The media will pick up this kind of thing and use it as a stick.'

Late in the day, the contingent of Dutch scientists at the meeting submitted a written statement suggesting fresh wording. They too were worried, it seemed, that the 'best guess' might be interpreted as something more concrete than it was meant to be, blinding politicians to the fact that it might prove to be an underestimate. 'Despite many uncertainties,' they wrote, 'we are concerned about our finding that future rates of climate change may exceed any rate of change ever experienced by humankind in the past. There are no reasons to expect that humankind itself, or the ecosystems on whose functioning humankind depends, will be able to adapt to such rates of change. A further point of great concern is that, although we have confidence in the results of our assessment, the complexity of the system may give rise to surprises. The prime example of such a surprise is the totally unexpected appearance of the ozone hole, notwithstanding many previous assessments of the state of knowledge of the ozone layer.'

The reference to the 'scope for surprises' made it into the report. But without elaboration.

I left the meeting with mixed feelings. On the one hand, I was dissatisfied that the scientists, as a group, had pulled their punches on spelling out what they thought was the worst case. I knew from private conversations that many of them considered amplifying feedbacks to be a huge danger. Yet they couldn't bring themselves to spell this out graphically in the report, which was going to provide the basis for negotiations by well over a hundred governments. In any threat assessment involving military security, the worst-case analysis would always be considered. Indeed, most governments would probably end up basing their defence policies on it: buying perpetual multibillion dollar insurance against invasion in the form of military procurement. Why should this environmental security threat assessment be any different? After all, global warming had the potential to seize territory and lay waste to land no less efficiently than an invading army. Given time, sea-level rise, drought, flood and pestilence could do that job just as well as tanks and bombers.

Nevertheless, I knew that the world had been provided with a warning on global warming that would be difficult to ignore. Sooner or later, governments and industry progressives were now going to have to do something about greenhouse-gas emissions, that was clear. If Margaret Thatcher could react the way she had in accepting the threat, other world leaders might do the same.

The oil and coal industries, and their dependants such as the automobile and electric utility industries, now had a big problem. The IPCC scientists had spelt out clearly in their report what would be needed to stop the inexorable increase of greenhouse gases in the atmosphere: deep cuts in emissions. This applied in particular to carbon dioxide. The passage in the report that would clearly be the most quoted in the years ahead pointed out that if atmospheric concentrations of carbon dioxide were to be stabilized at present-day levels, themselves higher than they had ever been for at least 160,000 years, emissions would need to be cut by 60 per cent or more – immediately. The longer the world delayed these cuts, the deeper they would need to be.

Entering the 1990s, the oil industry had enjoyed more than a century of hegemony. Yet in over a hundred years of oil burning, we were still not halfway through all the oil ever found. We had burned less than 700 billion barrels, with well over 1,000 billion barrels mapped out below ground ready to be pumped up and used. To that could be added all the oil yet to be found, perhaps another 700 billion barrels according to industry estimates at the time. At an annual burn rate of some 22 billion barrels, the oil industry clearly had enough reserves, and as-yet undiscovered deposits, to reasonably expect a repeat performance spanning most of the twenty-second century. How would the great companies created during the first oil century react to the prospect of being deprived of a second century?

For the coal industry, the arithmetic of carbon was still more daunting. By any definition of deep cuts in emissions in the decades to come, the vast majority of 8,000 billion tonnes of coal deposits then estimated to exist on the planet would have to remain below ground, unburned.

Like the tobacco industry before them when faced with evidence of the ruinous impact of their product, the choice for the carbon industries was stark: denial, obfuscation – and worse – on the one hand, and open embrace of a paradigm shift in their core business on the other.

AUGUST 1990, SUNDSVAAL, SWEDEN

With over four million paid-up supporters around the world in 1990, Greenpeace at the time had an annual budget of some $160 million, and with that sum – tiny though it was compared with the resources of the companies and governments it took on – it fought campaigns on an international basis, and actually won some of them. By 1990, the whale campaign had been instrumental in bringing about a moratorium on whaling, the Antarctic campaign was en route to having Antarctica declared off-limits for oil drilling and mining, and the ozone campaign was beginning to play a major role in setting up a permanent ban on CFCs. Such campaigns were based on a mix of the non-violent direct actions for which Greenpeace was so famous, and behind-the-scenes lobbying. Just as much effort went into the less glamorous work of corporate and governmental lobbying as was invested in harrying whalers, blocking toxic-waste pipes, and so on. My new colleagues wore suits and twinsets just as often as they did anoraks and wetsuits.

It was not long before I and my suit were called up for lobbying work – international duty at a vital plenary of the IPCC in the sleepy

Swedish town of Sundsvaal. During the summer, two other IPCC reports had emerged: one on the anticipated impacts of global warming, and the other on potential policy responses. At this full meeting of the IPCC in Sweden, governments had the task of combining these reports with the scientific report into an overall IPCC First Assessment Report. This would, in turn, be presented to a ministerial-level gathering of governments at the World Climate Conference in November 1990 at Geneva. It would be at this conference that the nations of the world would decide what to do about the threat of global warming.

Under the terms of the IPCC's mandate, the final report had to be a consensus document of all governments present in Sundsvaal. This was to be some task. In work completed by the IPCC as a whole, as of August 1990, the political geography was already clear. There was a yawning gulf between the scientific assessment report and the policy responses report. The IPCC's scientific working group had professed itself 'certain' that global warming lay ahead unless greenhouse-gas emissions were cut. The impacts working group had predicted a collage of expanding environmental catastrophe should the IPCC scientists' predictions turn out to be correct. But all that the policy responses working group had come up with, after 18 months of deliberation, was a toothless list of potential technologies which could help, in principle, with the limitation of greenhouse gases. This third working group was chaired by the United States.

All this reportage, and the massive requirements it imposed on the paper industry, might have seemed at first sight to be a recipe for death by boredom for governments. But the stakes were high. At least, the oil and coal industries clearly suspected as much. Among the dozens of non-government organizations registering alongside Greenpeace International for the IPCC plenary in Sweden were the Global Climate Coalition and the Global Climate Council. These sounded for all the world like scholarly and neutral Washington think-tanks, but a scratch below the surface revealed otherwise. The board membership of the Global Climate Coalition included representatives of the American Petroleum Institute, Amoco, Arco, Phillips, Texaco, DuPont and Dow Hydrocarbons. Shell and BP were members. The major users of oil were there too, in the shape of the Association of International Automobile Manufacturers and the Motor Vehicle Manufacturers

Association. Coal interests included the American Electric Power Service Corporation, the American Mining Congress, the Edison Electric Institute and the National Coal Association. The Global Climate Coalition was in fact the main umbrella organization for the oil, coal and auto industries' response to the global warming issue.

Nor did it operate alone. I had heard tales of the antics of a prominent industry lobbyist whose suggestions for watering down the policy-responses report had been even more brazen than those that Exxon's Brian Flannery and other carbon-fuel industry scientists had tried to impose on the scientists' group. This man's name was Don Pearlman. A former under-secretary in the Department of the Interior during the Reagan years, Pearlman now headed the second carbon-fuel front group, the Global Climate Council. He came from a Washington law firm, and his clients – the funders of his front organization – were, unlike those of the Global Climate Coalition, unknown because there was no requirement to register them.

On the first evening in Sundsvaal, as I entered the plushest hotel in town to attend a side-meeting of the scientists' working group, Don Pearlman was seated in the lobby with five diplomats, all Arab, including the head of the Saudi delegation. They had their heads down, copies of the draft negotiating text for the IPCC final report open in front of them. Pearlman was pointing at the text, and talking in a forceful growl. He looked like a professor holding a tutorial class. As I walked past, I saw him pointing to a particular paragraph and I heard him say, quite distinctly, 'if we can cut a deal here . . .'

At the time, although it seems so naive now, I was shocked.

Hundreds of negotiators lined the long tables, each team arrayed behind their country name-flag. At the back of the hall, the representatives of environmental groups and industry sat cramped together. Teams of interpreters, one for each UN language, huddled behind glass windows in booths along the side walls. Outside the hall, barred from the proceedings, dozens of journalists patrolled impatiently, waiting to pounce on delegates en route to the washrooms. The show was about to begin.

Not long into proceedings, the Federal Republic of Germany, as it was then, took the microphone, and with it, pole position in the negotiations. The West German government was convinced of the magnitude of the threat, said the head of the delegation, and had decided that it was going to cut carbon dioxide emissions, unilaterally, by 25 per cent of present-day levels by the year 2005.

This was a promising start. But soon after came the head of the Soviet delegation. We hadn't gathered here to pat each other on the back, he said. The figure that concerned him was the 60 per cent cut in carbon dioxide emissions mentioned by the IPCC's scientists as necessary to stabilize atmospheric concentrations. It appeared in the report only as an example, without comment as to whether it was advisable. Would the changes be dangerous or not, he wanted to know, if atmospheric concentrations continued to rise? He wasn't sure. He worked his way ponderously to a bottom line that was to become familiar. Did we, in fact, need to take any measures at all to cut emissions?

It was now clear where the Soviet Union's interests in climate change were going to lie: with Siberia, in its role as the new frontier for the international oil industry, and with Kazakhstan and the Caspian, in their roles as barely tapped oil provinces.

The interventions came and went as delegates from progressive nations digested this setback. Among them, Norway, claiming to speak for all of Scandinavia, had a pointed early observation to offer: 'We do not find it proper that non-experts in this field lay more emphasis on uncertainties than the experts themselves do.'

This opinion was not to be a deterrent to Saudi Arabia. Soon they took the floor to support the Soviet Union, and to claim that the scientific uncertainties were not reflected in the draft executive summary of the final report.

Then came Italy, speaking for the EC. 'We are an alarm-minded group of countries, Mr Chairman,' said the head of the Italian delegation. The EC, he went on, would be meeting to find a common position early in November, before the World Climate Conference later that month.

Having bided its time, the United States made its first intervention. Fred Bernthal of the State Department, head of the US delegation and chairman of the problematic policy-responses working group, sounded measured and reasonable. Yet his words struck a chill in many hearts in that hall. The draft, he said, stressed the negative impacts of climate change while passing over the positive benefits. Also, the policy-responses group had not had time to consider the vital social and economic costs of action to limit emissions. This should be stated clearly in the report. This was the first of many airings of the Bush administration's view that it would cost too much to cut emissions.

Bernthal had some final advice in his first intervention. 'All countries need to consider how they can adapt to climate change.'

The American environmentalists around me at the back of the hall grimaced. Theirs was going to be a long, tough campaign.

Canada was not to be deterred by her neighbour's bullishness. 'We wonder,' said their head of delegation, 'if the executive summary is crisp enough to reflect to people that the evidence is compelling.'

New Zealand joined Canada. There was not enough mention of the coral island nations, they said. The impacts of sea-level rise on them would be severe even at minimum projections.

Soon after, the IPCC heard the first words from an island nation. The Foreign Secretary of the Republic of Kiribati (pronounced 'Kiri-bass') had flown to Sundsvaal from an idyllic coral atoll in the centre of the Pacific Ocean. Peter Timeon was jet-lagged, culture-shocked and nervous. He began with the normal diplomatic pleasantries, and moved swiftly to Mrs Thatcher's reaction to the IPCC scientists' report. He read to the hall her words about the 'authoritative early warning' and its impacts on our daily lives.

Behind the UK's flag, John Houghton, chairman of the IPCC's scientists, turned round to see who was speaking.

'I need not remind this distinguished meeting of the nature of this authoritative early-warning system for us in the Republic of Kiribati and the countries of the South Pacific,' Timeon said. 'What the IPCC scientists' exhaustive analysis predicts will affect not just our daily lives in Kiribati, and the other low-lying atoll countries of the South Pacific, but the very future of our nations and the whole cultures of our people.'

Timeon continued, still audibly nervous. 'I make no apologies in noting what this authoritative early-warning system heralds in a world which does not make any effort to significantly cut its greenhouse-gas emissions. It does mean, Mr Chairman, that within only a few decades, average surface temperatures on our planet will be higher than during any period of humankind's tenure on planet Earth. It will therefore, Mr Chairman, be totally irresponsible of us to wilfully bequeath our children this kind of future without any effort to avoid it.'

At the US desk, Bernthal's face was impassive. The head of the Saudi delegation wore a faint frown. Pearlman's jowled face was unreadable.

Timeon was now warming to his oratory. 'Mr Chairman, distinguished delegates. I wonder how many people in this hall know where my country is? Without wishing to take up so much of this meeting's precious time, it suffices to say briefly that there is no place among its 33 tiny atolls that rises higher than two metres above sea level. Long before the sea level rises that far, my country and others like it will have been condemned to annihilation.'

Timeon had read the draft overview with great concern, he said. 'Mr Chairman, that document fails to recommend policy responses commensurate with the dire warnings of the IPCC scientists, and as such it fails to offer the low-lying island nations any prospect of escaping the greenhouse threat.'

The 15 independent island member-nations of the South Pacific Forum, at their annual meeting earlier that year, had issued a communiqué recognizing the extent of the threat, and what was needed to meet it. Encouragingly, both Australia and New Zealand were signatories to it. It noted, furthermore, that most of the problem came from the North. It therefore strongly urged industrialized countries to enforce significant cuts in the emissions of greenhouse gases, and to establish obligatory emissions-reduction standards.

The hall was quiet. Perhaps a hundred heads among the several hundred diplomats and scientists were now craning round at the Pacific islander.

'Mr Chairman, it is quite obvious that to escape from the threat of global warming, a concerted international action is needed to drastically decrease our consumption of fossil fuels. The time to start is now. In the low-lying nations, the threat to us of global warming and sea-level rise is *frightening*.' He laid emphasis on the word, and paused.

'Mr Chairman and distinguished delegates, I hope this meeting will not fail us. Thank you.'

A wave of applause swept the hall.

No other statement had elicited so much as a flicker of recognition from the listening masses. Clapping was just not the way of things at UN meetings. In the two years of negotiations that were to come before the Earth Summit, I very rarely saw a statement from the floor spontaneously applauded in that way. From the back, I estimated that at least half the assembled diplomats were applauding Timeon.

He looked around the hall, his bright Micronesian eyes sparkling, the beginnings of a smile playing on his lips.

Peter Timeon became the darling of the press corps in Sundsvaal. In the press room, newspaper articles about him, and about the threatened island communities for which he was *de facto* ambassador, were plastered across the bulletin boards.

But he might as well have stayed at home as far as the White House was concerned. The US delegation tabled a catalogue of attempted emasculations of the text. The US, Saudi and Soviet delegations – a strange new trio of bedfellows on the UN scene – chipped away at the draft, watering down the sense of alarm in the wording, beefing up the aura of uncertainty. European countries led the attempts to stick to the text agreed by the scientists' working group. For a long period, as the science section of the overall report was discussed, the Saudis made one intervention in about every five, often trying to change text agreed long before by the IPCC working groups. Don Pearlman, sitting at the back of the hall with his collaborators from the Global Climate Coalition, wrote it all down in his notebook. I wondered if he was paid bonuses by his mystery paymasters for each intervention.

The environmentalists had by now invented a name for the foot soldiers for the fossil-fuel industries: Pearlman, Flannery, the Global Climate Coalition people and their hangers-on. We called them the carbon club.

The question of how to describe the impacts of climate change brought the political games to an unpleasant focus. The draft text, already a compromise, read 'climate change will bring both positive and negative impacts, although the negative ones will dominate.'

The Soviet head of delegation renewed his effort to muddy the waters. Agriculture in many areas may improve, he said. Among others, this provoked France. 'I don't give a hoot,' said the French head of delegation. 'If my country is about to be wiped off the face of the planet, it won't matter to me if agriculture is boosted somewhere. We should stress regional negative effects.'

John Houghton proposed a new sentence: 'Rates of change will bring negative effects.' This, he said, was the crux of the matter if one looked at the ecology of it all.

Then the United States. No, said Bernthal, the USA agreed with the USSR. This language should go.

Truly, the Cold War was over.

The next day, Mustafa Tolba, head of the United Nations Environment Programme, met a group of environmentalists. I was among them. Tolba had been the architect of the Vienna Convention on Protection of the Ozone Layer, and the Montreal Protocol to that convention, in which governments had committed themselves to cuts in the CFC gases which cause ozone depletion. It had taken 14 years to make governments define that plan.

It was people, scared people, who had forced governments to act, Tolba told us. But in the case of ozone depletion, the 'proof' that CFCs were eroding the ozone layer had been as good as you could get. In 1986, NASA had flown a converted spy plane through the Antarctic stratosphere and found the 'smoking gun' of chlorine which could only have got there from CFCs. And where the chlorine was high, the ozone was proportionately low. With global warming, the genuine scientific uncertainties would give plenty of opportunity for obfuscation. Tolba was very direct with us. We had a vital role to play, he said. But we should not underestimate what we were up against. 'The alliance between oil and auto companies,' he said, 'is one of the most powerful alliances in the world. It can paralyse governments.'

Back in the negotiating hall, it was doing just that. The US delegation had spent an hour and a half trying to get the words 'climate change' replaced with the words 'global warming at the surface of the Earth'. There was little proof as yet, they said, that climate change would result from this warming.

The suggestion was met with barely suppressed annoyance, and, from Austria, ridicule. 'Maybe we should change the name of this panel,' they said, 'to the "Intergovernmental Panel on Global Warming at the Surface of the Earth".'

Wry laughter filled the hall.

We entered the final day. A new draft of the negotiating text had been prepared overnight by the secretariat. The wording in the original science report which described the degree of certainty scientists had in their prediction of global warming, in a business-as-usual world, had been strong. 'We calculate with confidence,' it read. That was the form of words replicated in the new draft.

The USA and the Saudis duly objected. They wanted a compromise form of words that dropped the word 'confidence'.

'I've heard the word "compromise" several times,' said the head of the Swiss delegation. 'You can't compromise with science.'

And so it went on.

As evening turned into night, deliberations centred on wording which gave more emphasis to the need to understand the costs of climate change. The Saudis preferred 'costs and benefits'. The US supported them. The atmosphere became edgy as the number of countries waiting to intervene backed up.

Saudi Arabia was willing to delete text to make room for its suggested additions. It suggested deleting 'an important aim would be to define targets for carbon dioxide and other greenhouse gases'. This brought a burst of derisory laughter from the floor.

As the clock turned, the Saudis upped the ante still further. They suggested replacing the words 'carbon dioxide' with the words 'greenhouse gases' throughout the report. The laughter which greeted this was loud and long, and the Saudi head of delegation looked flustered. I studied Pearlman, writing in his book near me. His face gave nothing away. Was this the deal I had overheard him angling for in his tutorial for OPEC negotiators?

The chairman, eminent Swedish scientist Professor Bert Bolin, seized the opportunity. 'Let's not beat about the bush,' he said. 'We all know why we are here. Will you withdraw the wording?'

The Saudi head of delegation hesitated, frowning at his microphone. This was the moment of choice for him. Stop blocking and go along with the consensus, or quit the whole process, losing the chance to stall and water down text in the treaty negotiations that now looked likely after the World Climate Conference.

'For the sake of compromise, we will,' he said slowly.

Pearlman sat impassively amid the Global Climate Coalition's lobbyists as the final wording of the IPCC's First Assessment Report was agreed unanimously by over a hundred governments.

The clock said 4 a.m.

OCTOBER 1990, TARAWA ATOLL, REPUBLIC OF KIRIBATI

In the run-up to the World Climate Conference, the carbon club's foot soldiers were spread about the world, lobbying ministers in any country of consequence, while no doubt concentrating their core preparations on ministers and officials in OPEC capitals. Greenpeace had allocated me the entire Pacific, or as much of it as I could cover in the time available. During September and October, accompanied most of the way by my Samoan Greenpeace Pacific colleague and great friend, Pene Lefale, I completed a circuit of meetings with government ministers in Tokyo, Sydney, Canberra, Auckland, Wellington, Fiji and Kiribati.

In his office on the atoll of Tarawa, the President of Kiribati, Ieremiah Tabai, greeted Pene and me in shorts and flip-flops. French windows behind him opened directly onto a ribbon of white coral sand shaded by palms. The president's desk faced out across kingfisher-blue shallows shimmering before a royal-blue vastness beyond the reef. Tarawa's necklace of islands, rarely more than 200 metres across, wrapped around a 15-kilometre-wide lagoon. In that tepid patchwork quilt of azure and aquamarine I had earlier in the day swum and snorkelled with Pene. We were in paradise.

Unsurprisingly, President Tabai wanted to discuss the IPCC's sea-level rise projections. The best-guess estimate in the IPCC Report was of a 30-centimetre increase in average sea level over the next 50 years, assuming business-as-usual emissions. In a land where fresh water was everything, this seemingly small rise would spell utter calamity. The coconut palms and breadfruit trees that draped the islands depended on a lens of fresh groundwater, recharged by rainfall, sitting below the coral sand and gravel: a permanent but fragile life-giving bubble sandwiched between the salty worlds of ocean and lagoon. In the villages through which I had walked since my arrival, I had seen simple wells excavated into the top of the two-metre thick lens, just 15 metres or so from the shore at high tide. The water in them was cool, fresh and clear. It sustained all life on the islands, plant and animal, directly or indirectly. The people of Kiribati supplemented their fish and coconut-based diet with babai, for example, a plant grown in flooded pits cut into the groundwater lens.

It was difficult to imagine a way of life more vulnerable to the sea. Long before the rising sea had eroded their living space, the people of Kiribati would be facing massive saline invasion of their groundwater. And this peaceful, sustainable, centuries-old civilization was but one of dozens of Pacific cultures threatened in this way.

At the South Pacific Forum meeting earlier in the year, President Tabai had exhorted his fellow Pacific heads of state to form a united front at the World Climate Conference and beyond, and to plead for a rapid reduction in emissions from the industrialized nations. All 15 leaders had supported this view, including, critically, the Australian and New Zealand Prime Ministers. It was vital, Tabai now told us, that Australia and New Zealand continued to show that kind of solidarity with their Pacific neighbours.

'The problem is how to persuade the industrialized nations,' Tabai said, looking out across the ocean. 'If the Pacific nations speak as one voice, maybe we can help change the world's dependence on oil and the rest. There is still time. For us there is no choice.'

NOVEMBER 1990, GENEVA

Less than a month later, I sat with several dozen colleagues in a room in a drab hotel by Geneva's central railway station. An air of excitement hung over us. Interest in the World Climate Conference, beginning the next morning, was running high around the world. We knew that media registrations for the event had run into many hundreds.

Heading our delegation was Paul Hohnen, a former Australian diplomat. Hohnen's road to Damascus had been similar to my own. Once an ambitious member of his country's foreign service, he had risen meteorically to become an acting high commissioner. Then one day he had woken up to find that he was no longer able to let his daughters play in the sun in his Canberra backyard. Something called an ozone hole was drifting over them, out of the freezing skies above Antarctica. The life-protecting ozone layer was being depleted by CFC gases produced in millions of tonnes each year by chemical companies. That had been too much for Hohnen: he wrote to Greenpeace offering his services. In blazer, striped shirt and regimental tie, he looked about as much a traditional environmentalist as George Bush.

Hohnen now laid out his plan for the Greenpeace delegation, which consisted of a small international team inside the summit, and a much larger team on the outside. The latter, a group of informally dressed Germans and Swiss, listened in as Hohnen ran through his intelligence-gathering on government positions and intentions. The outside team would have the more dramatic role in Geneva: the execution of whatever non-violent action, or actions, we decided to carry out during the summit.

Over the first few days of the World Climate Conference, more than 700 scientists from around the world were due to meet to review the technical aspects of the IPCC Report. The governments wanted them to negotiate a Scientists' Declaration, summarizing the nature and magnitude of the greenhouse threat. Simultaneously, government officials would be negotiating a Ministerial Declaration. This was to be the all-important product of the conference: it would be finalized, and signed, when environment ministers attended the last few days of the event. It would effectively be the terms of reference for full-scale negotiations for a treaty, the Framework Convention on Climate Change. Such negotiations, it was commonly accepted, were now inevitable in 1991. The most important thing about this ministerial agreement, therefore, would be whether or not it mentioned targets and timetables for cuts in greenhouse-gas emissions.

Entering the World Climate Conference, every industrialized country save the USA had set itself unilateral targets for at least putting a cap on emissions at present-day levels by the year 2000. Some, led by Germany, had pledged to cut emissions.

On Sunday 4 November, the Scientists' Declaration was made available to the press. 'A clear scientific consensus has emerged,' it announced, 'on estimates of the range of global warming that can be expected during the 21st century.' The statement gave a clear conclusion for governments as to what this all meant. 'Countries are urged to take immediate actions to control the risks of climate change.'

Urged. A strong word for scientists.

The next day, the final day of the World Climate Conference, Greenpeace brought out its demonstrators. By mid-morning, the main entrance to the UN's Palais des Congrès, and its side entrance, were ringed by Swiss and German citizens, chained together. The blockade was total, if symbolic. Diplomats came and went, but had to duck under the chains. A barrage balloon floated in the sky over the Palais, urging delegates to 'Cut CO_2 Now'. Boiler-suited Greenpeace activists swarmed over the roofs of the Palais, clutching placards bearing the same message.

From inside the building I watched the demonstration through the windows. The riot police were out in force, but merely watched from the sidelines. The reason for this uncharacteristic leniency, we learned, was that the Swiss Government was trying to introduce a carbon tax at the time, and wanted to be seen to be under citizen pressure.

While the demonstrators braved the icy weather outside the Palais, the diplomatic ducking and weaving continued inside. The USA was the only major country not to send a representative of ministerial status. The head of the US Environmental Protection Agency had stayed at home. His press secretary had told the Reuters press agency that he didn't feel the summit required his time. The job of holding the fort was given to Dr John Knauss, a diffident man in a bow tie described by the *Chicago Tribune* as 'a lower-level official from the Commerce Department'.

Knauss had to submit himself to a full press conference. It was packed, and hostile. Did the US delegation deny that there was any sort of problem, one journalist asked. Knauss, to his credit, couldn't swallow the full extent of his White House brief. 'I don't know of a single credible scientist,' he replied, 'who doubts that if we put greenhouse gases into the atmosphere at rates we are today, there will be an effect on climate. The only uncertainty is the degree of warming.'

The questions came at him thick and fast. 'Isn't it true,' asked UK's Channel 4 news, 'that the US, the USSR and the Saudis are in a coalition of interests slowing progress?'

Knauss waffled, his manner uncertain. He couldn't bring himself to issue an outright denial. 'None of you were in the sessions,' he said finally.

That had to be my cue. 'Those of us who *were* in the sessions can report that there was traffic between the US, Saudi and Soviet desks, and that it involved American oil and coal industry lobbyists as well as delegates. Would you not agree that this sends a cynical and irresponsible message to the world?'

Knauss glared at me. 'No.'

My mistake. Never present the opportunity of a simple yes or no.

Then came a young man in a T-shirt. He was even more of an unaccredited journalist than me. 'I am from the United Nations Association youth branch, and I represent the young people of the world,' he said. 'I'm not going to ask you a question, since you haven't really answered any yet. I'm going to give you a petition signed by 17,000 young people, reminding you that you did not inherit the Earth from your ancestors, you borrowed it from me and millions like me around the world. And urging you to act.'

He threaded his way through the crowd to where Knauss sat. The American head of delegation took the document, his face a twisted mask. Most of the journalists at the press conference clapped.

Back in the main hall of the Palais des Congrès, the closing negotiations on the Ministerial Declaration were under way in a packed plenary session attended by the ministers themselves. Diplomats negotiating declarations and treaties leave undecided text 'bracketed'. A group of ministers led by Europeans, in a behind-closed-doors session the night before, had agreed ways to deal with the numerous phrases in the ministerial declaration which still had brackets round them. The compromise form of words appeased the USA by not advocating commitments to cut or freeze emissions, and allowed the Europeans to note that 'some countries' had made 'decisions and commitments aimed at stabilizing their emissions of carbon dioxide'. The crucial words, 'cuts' and 'targets', were nowhere to be seen. The small island states, to whom in particular this omission meant so much, had not been invited to the closed-door session.

Earlier in the day, a few of the island nations had made history by forming a totally new bloc in the UN: the Alliance of Small Island States. Caribbean and Pacific governments were joining in droves. The Alliance, AOSIS as it was universally to become known, would soon grow to include more than thirty island countries around the world. This final session of the World Climate Conference was to be their first taste of working together. Most of these governments now knew, like President Tabai of Kiribati, that they had no margin for error, and no place to go.

Two brilliant young environmental lawyers had been heavily involved in the creation of AOSIS. Phillipe Sands and James Cameron were passionate advocates of action on global warming, and sought to engineer opportunities to use international law to protect the atmosphere. Together with Caribbean and Pacific diplomats, they quickly drafted the founding principles for AOSIS in Geneva. These principles stressed a key concept in the environmental debate of the day, the so-called precautionary principle, which held that in the absence of complete scientific certainty about an environmental threat, policy would be predicated on the need to act ahead of absolute proof. As far as environmentalists were concerned, the precautionary principle applied to many issues being debated at the time. In the case of the enhanced greenhouse threat, it clearly meant setting targets and timetables for emissions reductions based on the scientific information already on the table, not waiting for more. To wait for the IPCC to bring in final irrefutable proof – a 'smoking gun' such as atmospheric scientists had found with the Antarctic ozone hole – might mean waiting until environmental and economic disaster had descended. And such disaster would strike the island states before anyone else.

AOSIS had the chance to air their grievances about the absence of the all-important reference to cuts and targets in the draft World Climate Conference Declaration when Argentina insisted that specific mention of carbon dioxide should be made more frequently in the text of the declaration. A string of island countries from the Caribbean and Pacific queued up to agree.

'Is there anyone in the room who opposes the Argentinian proposal?' the French chairman asked, looking around the great hall.

Saudi Arabia did. 'If we open those brackets we agreed,' the Saudi said, referring to the behind-closed-doors deal, 'then we will open all brackets.'

The Italian Environment Minister stepped in. He suggested mentioning 'carbon dioxide, methane, ozone precursors and other greenhouse gases'. That would serve to take the focus off carbon dioxide, which was what both the Saudis and the Americans had fought for throughout. Among other things, it would entail equal status for the carbon dioxide emissions of Cadillacs in Hollywood and Riyadh, and the methane emissions of subsistence rice farmers in the developing world.

'For the sake of compromise, yes,' said Saudi Arabia. 'But then we should drop mention of carbon dioxide in other paragraphs.'

Ironic laughter rippled around the hall.

A charismatic environment minister from Trinidad and Tobago, Lincoln Myers, had been a key player in the creation of AOSIS. He was now champing at the bit. 'The extent to which politicians are quibbling while the planet is going to pot is very worrying,' he said. He had a deep, statesmanlike speaking voice, with a lilting Caribbean accent. He separated his words for emphasis. 'I do not want one or two men or countries to decide what my fate is.'

Brazilian Minister of Science José Goldemberg joined the attack. 'Energy is carbon dioxide. We can't mention methane and other greenhouse gases in the same breath,' he insisted. 'We can't change science to please the producers of coal and oil.' Then came Zambia. 'We have demonstrators outside here telling us to listen to the destruction of our climate and our nature. Telling us to listen.' They were still all there, chained together in the freezing late afternoon as the sun set over Geneva.

Now John Knauss took the floor for the USA. As a sometime scientist, he said, he wished to make three points. First, methane was indeed an energy gas. Second, its concentrations were rising at twice the rate of carbon dioxide. Third, nitrous oxide was an important gas coming from energy too.

Argentina differed. Although they were exporters of oil, they were suggesting doing something contrary to their own interests in wishing to keep the reference to carbon dioxide as it was. Why could others not do the same?

Italy now made another suggested compromise: 'carbon dioxide and other greenhouse gases such as methane, nitrous oxide, and ozone precursors'.

Did Argentina agree? asked the chairman.

Yes.

Did Saudi Arabia agree?

Yes.

Applause filled the hall. Many of the diplomats were thinking of evening planes out of Geneva.

Kiribati raised its flag. The minister did not wish to take much time, he said, but in the sentence that referred to the threat to the island countries, he simply wished to make the wording more accurate: the greenhouse effect and sea-level rise 'would' threaten survival, not 'could'.

Are there any objections? asked the chairman.

Knauss flashed up the US flag. I was told much later, by someone in a position to know, that John Knauss was a decent man. He certainly gave little indication of it that day. The declaration had undergone negotiation over many long evenings, he said. The United States opposed any change made at this stage.

The next speaker was one of those thinking about his plane. Only 13 minutes remained before the conference was due to close, he complained. I did not see where he came from. He agreed with the US: 'We will be here for days if we discuss matters which are not of substance.'

The Minister from Kiribati spoke quietly, no sign of emotion in his voice. 'I'm sorry that some think of it not as a matter of substance. We think of it as a matter of life and death. The quote from the scientists' document is "would", not "could". But perhaps I will go along with the change, in the interests of compromise.'

Italy now suggested that, to save time, the whole document, save the few remaining bracketed sections, be immediately adopted.

But no. Now Western Samoa insisted on taking the floor. 'I have come a long way,' said the Samoan Minister. 'Here, we are a long way from the South Pacific. I want to come back to Kiribati's point. Why compromise? Sea-level rise would threaten our survival!'

The chairman tried to kill the proposal, and asked Western Samoa to step down. The Samoan minister insisted that the question be returned to Kiribati. The Minister from Kiribati was conciliatory. He had come from the other side of the world, he said, to agree matters of life and death. But he would now agree to this compromise. The chairman, looking hopefully around the hall, asked if he could now consider the articles not in brackets accepted. Clapping filled the hall.

But two more flags had gone up: the Republic of Nauru, that tiniest of Pacific island nations; and Trinidad and Tobago again.

The delegate from Nauru did not want to add to the delay either, he said, but the chairman had mentioned earlier that comments could be put in the minutes. The Republic of Nauru wanted the fact that it supported the Republic of Kiribati put in the minutes.

Trinidad and Tobago had a completely new issue. Minister Myers had said earlier that he wanted to raise an amendment. He would not be dismissed as a codicil, he said. His proposal was to include the statement that 'all measures to meet the climate challenge must be based on the precautionary principle.'

'But we agreed unanimously, by clapping, to accept the document,' cried the next interventionist.

Now Barbados. 'Trinidad and Tobago did ask before for the floor. It is a valid comment. We cannot be treated like a codicil to this monster. We cannot be swept aside by applause, like at a football match.'

This itself was met with applause.

And so they plunged back in again. Most governments accepted the precautionary principle, India observed. But some didn't. The delegates needed to go home sometime. He appealed to the island nations not to insist on this today. France agreed. So did Austria, but with a subversive sting in the tail, taken from Nauru's lead. 'We will adopt this statement as it is, but only for the sake of compromise, and reluctantly. We want to put on the record that this is an imperfect and insufficient compromise. Other countries may wish to join us.'

Lincoln Myers now stirred the pot once more. 'Some indicate that they wish to go home soon. If we continue making declarations like this one, in a few years' time some of us will have no homes to go to. In the circumstances, we have got to base actions on the precautionary principle.' Having said that, though, he conceded that he would withdraw this amendment for the time being.

The chairman thanked him, relief clear in his voice. His comments would be recorded in the minutes. Were there any others?

It now became clear that Nauru and Austria, between them, had unlocked the floodgates of discontent. The statements came fast and furious, and with passion. First came Bangladesh, disappointed that its special plight had not been recognized or catered for. Then Mauritius, wishing to support Kiribati. Then St Lucia, and Barbados, supporting Kiribati and Trinidad and Tobago. 'We are not the polluters,' said Barbados, 'we are the first line of victims.' Then the Sudan, worried about desertification, and supporting Bangladesh. Then the Cook Islands, Afghanistan and the Maldives. Then Tuvalu and Western Samoa.

Knauss sat slumped behind his flag throughout all this. I wondered if he was feeling guilt at his country's stance in the face of this display of anguish from the small island states. It was Saudi Arabia who asked for proceedings to be cut off. Indonesia backed them. The Solomon Islands squeezed in one more entry for the minutes, supporting Kiribati and Trinidad and Tobago.

Let there be no more statements, the Chairman pleaded. 'The Declaration is submitted for adoption.'

And the flags stayed down.

The governments of the world had put themselves on course for long-running negotiations on the limitation of greenhouse-gas emissions. The next time they were to discuss global warming, they would be negotiating a treaty. Even in the absence of a commitment to agree targets and timetables, the threat to fossil-fuel interests had now assumed sufficient official recognition to warrant the creation of a permanent international forum. And maybe, just maybe, the spirit of the small island states – backed as it now clearly was by many progressives in the governments of developed countries – might one day find a way of translating itself into commitments to emissions cuts deep enough to stave off the threat of global warming.

The carbon club, at any rate, now knew for sure that it had a potential problem.

2

The Price of Oil

January–December 1991

FEBRUARY 1991, CHANTILLY, VIRGINIA

The year 1990 was the hottest since records began. But by the time the negotiations for a convention on global climate change opened, the global warming issue itself had cooled right down. On the Saturday night before the Monday start of the first session of talks, I lay half-awake in a hotel in Washington watching on CNN the flickering lights of a Scud attack on Saudi Arabia.

I had nothing to kill except Sunday morning, which I spent walking around the city. Opposite the White House, a rank of police cars lined the middle of the road. Within the fence, secret servicemen patrolled the lawns. The object of their concern was a desultory peace demonstration on the other side of the road. Fifty or so demonstrators had established a makeshift camp on the grass. Behind ragged banners, half a dozen drummers kept up a constant beat. Cars rolled past on Massachusetts Avenue, yellow ribbons adorning their radio aerials. Hooting drivers held fingers up at the demonstrators, shouting derision. Around the corner, vendors were selling Operation Desert Shield sweatshirts by the boxful. The slogan of the hour was emblazoned on them: 'Support US troops for World Peace'. At a cost of a billion dollars a day, so the estimate went.

The next day, I rode out of Washington into Virginia. President Bush had moved the negotiations out of the capital and into the backwoods. The talks were to be housed in a palatial red-brick conference centre within musket range of the civil-war battlefield at Manassas.

After the first session of speeches, the Alliance of Small Island States held a press conference. By that time, only 15 journalists had registered. We were only 45 minutes out of the capital, but we might just have well been in the Sahara. Journalists were not about to leave their studios or desks in Washington, at this time, for a polite discussion about a bit of warm weather – even if it did involve more than a hundred governments.

The US delegation had come to the talks armed with a glossy booklet printed specially for the occasion, bearing the presidential seal on the front. US greenhouse-gas emissions in 2000 would be at or below 1987 levels, it announced. President Bush's men referred to the basis for this claim as the 'comprehensive approach'. The USA might be cutting greenhouse-gas emissions overall, but it would be *increasing* carbon dioxide emissions. Environmentalists called this ploy double-counting. The US would have to get rid of the CFCs anyway, of course, under the existing Montreal Protocol. The number one emitter of the number one greenhouse gas would still be hiking its emissions of carbon dioxide.

The opening speeches repeated the themes from the World Climate Conference, and with so many governments present these extended into the third day. Kuwait's turn came. Iraq's pillaging was not just of Kuwait, but of the environment, said the Kuwaiti Ambassador. Gone now would be the hundreds of species of plants in the desert, and the trees which the Kuwaitis had planted to stem desertification. Oil fires were turning the rain black, and putting 'dangerous gases' into the atmosphere. Not lingering on the shaky ground of atmospheric pollution, he turned to the oil slicks in the Gulf. His speech became a long, and heartfelt, diatribe against Saddam Hussein's crimes – environmental and human. Finally came a token allusion to climate, the reason the Kuwaiti was supposed to be here. 'There are many remaining uncertainties. These should not be overlooked in a rush to judgement.'

When the countries had finally finished their statements of position, the non-governmental organizations had their turn at the microphone. The Global Climate Coalition gave a speech on scientific uncertainties. Both the industry groups and the US delegation had taken to referring to 'potential' climate change at this point in the proceedings. They also referred to 'no regrets' policies, by which they meant policies that didn't cost anything. For example, if it didn't cost a company anything to retrofit its buildings with energy-efficient devices, when the accounting was done, then that was a 'no regrets' approach. For environmentalists, this was a frustrating reversal of meaning. A genuine 'no regrets' policy would be to cut emissions deeply, in case not doing so ended up literally costing the earth. Similarly, a truly 'comprehensive approach' would be to cut emissions of each and every greenhouse gas, not just one.

The negotiators finally broke into private session to discuss how many working groups they wanted to divide the talks into, and on what subjects. I went with my colleagues to the battlefield nearby. More than twenty thousand men had died there, American fighting American. I stood in the fields where so many had perished for slavery, thinking of those who would soon be dying for oil. On the houses around the Virginia site, Yankee flags and yellow ribbons now fluttered, emblems of how things had changed in the century that followed the American Civil War.

The first oil had been drilled in 1859, in Pennsylvania, only a few years before the Civil War began. The oil era sprang from the ashes with astonishing speed after the war ended. John D. Rockefeller, in many ways the father of the oil industry, had bought his first oil refinery in 1865, at the age of 26. Within 15 years he had made his first billion dollars, and was doing more business outside the United States than at home. He founded Standard Oil, the forerunner of Exxon, in 1870, which makes that particular oil giant over a century and quarter old. The other oil companies were not much younger, and no less instantly international in their reach. All the majors – Exxon, Shell, BP, Mobil, Chevron, Texaco, Gulf – were operating globally as fully integrated companies, self-sufficient in everything from production through refining, transportation and sales, by the time of the First World War.

That day on the civil-war battlefield, the task of achieving deep cuts in the burning of oil and the other fossil fuels should perhaps have seemed close to impossible. But then, as now, I felt a small but steady flame of hope. Had these powerful companies always experienced the unimpeded success they were enjoying in 1991? I knew that they had not. I had done my reading on the history of the oil era. I knew that the companies had faced four great threats to their short-term interests, before the current environmental one. The companies had seen each threat off by working together, either severally or as an essentially industry-wide cartel. But in doing so, they had flirted with disaster on more than one occasion. They were not invulnerable.

The first threat had been the issue of anti-trust legislation. It seems scarcely credible to people unfamiliar with the history of the oil industry, but governments in the first half of the oil century tried fairly regularly to break the cartel status of the great oil companies, and divest them into smaller competitive entities. Nowhere was this truer than in the United States. In fact, in 1911, a government once succeeded. The administration of Theodore Roosevelt, backed by the Supreme Court, used anti-trust legislation to force Exxon's forerunner to undergo complete divestiture. Public pressure was the key. Standard Oil's ruthlessly enforced monopoly, coupled with exposures of a web of illegal espionage and bribery, had elicited profound public disgust with Rockefeller and his creation.

As it happened, Standard/Exxon's 1911 divestiture proved only a temporary setback. The daughters of the divested company all soon became giant cartelized monopolists in their own right, showing scant respect for anti-trust laws. By 1923, for example, the chiefs of Shell, BP and Exxon were meeting in secret to carve up the international oil markets for their own optimal profit, fixing prices and quotas, all without any consideration of wider interests, whether national or international. By the 1970s, of course, the major oil companies had effectively become a separate government. When producer countries began calling the shots over pricing, it was the companies, not Western governments, that negotiated with them. Such was their propensity for collective action to protect their common interests that the companies became widely known as the Seven Sisters.

The second great threat was nationalization, again defeated by the ability and willingness of the companies to work together. For example, when Iran tried to nationalize oil companies working in the country in the early 1950s, the companies ganged up and blocked the use of Iranian oil. The outbreak of Iranian nationalism did not last long. The populist leader of the day was toppled in a CIA-backed coup. The Shah replaced him, and the rest is history.

The third threat was only initially perceived as such. It was the emergence of unity among the oil-producing nations. OPEC, once unified, challenged the companies' cartel with great success. The architect of OPEC's emergence as a power in the early 1970s was the then Saudi oil minister, Sheikh Yamani. Yamani did not call his strategy nationalization, but 'participation'. He sought, and got, what he called 'an indissoluble marriage' with the oil majors. The oil companies at first fought this tooth and nail, backed by bullying from parent governments. In 1973, US Secretary of Defense James Schlesinger actually threatened to invade Saudi Arabia. How interesting that looks juxtaposed with Saddam Hussein's antics in 1991. The Saudis actually had the oilfields wired to be blown up, in the event that rhetoric became reality.

When it became clear that Arab unity could not be broken, whether by diplomacy or by threats, the oil companies acceded to their changed status: they entered the indissoluble marriage. They then found that married life was not too unprofitable after all. The oil industry cartel had merely been replaced by a wider cartel involving both producers and companies.

In 1973, during the first oil embargo, the oil companies were placed in the position of having to administer the wishes of their new partners. Consumers in the USA witnessed the spectacle of transnational companies they had thought of as American administering a strict embargo on behalf of a foreign government. Hatred of the companies erupted across the country, and elsewhere in the Western world. This was a time when the phrase 'obscene profits' entered regular use, and when Americans took to wearing lapel badges exhorting 'Impeach Nixxon'. Again, an outbreak of public disquiet almost ended in disaster for the companies. At one point, 45 out of 100 US senators voted for divestiture of the oil giants.

All this had an unnerving effect in some quarters of the oil industry. Then the fourth great threat emerged: the extremely high oil prices engendered by the second oil shock in 1979. These stimulated strong government support for solar energy and other renewables for a while. Some companies began diversifying away from oil, developing strong interests in industries including mining, coal and chemicals. As this period of history shows, the oil companies do not have to remain wedded to their core business by universal decree. Many of the companies even bought solar companies at that time, though most sold them as soon as the Reagan/Thatcher years put paid to government support for renewable energy.

The fifth and final great threat is the current one. It has descended on the oil companies from a wholly new direction, and it strikes at the oil companies' product itself. Burning oil is the number one source of the number one greenhouse gas, carbon dioxide. And it is not as though global warming is the only problem. The *Exxon Valdez* spill in 1989 was a painful reminder that oil is a poison, a mix of some of the most carcinogenic chemicals known, a substance not only killing telegenic animals, but finding its way into food that humans would eat, and water that they would drink. Then there is the question of the air people breathe. Burning oil takes its toll in lung cancer, asthma, and all the rest. Since *Exxon Valdez*, American politicians have felt obliged to adopt two legislative nightmares for the industry, the Oil Pollution Act and the Clean Air Act.

How will the oil industry ultimately deal with this new threat? Will it deploy the traditional methods it used in dealing with its other major threats? Will it resort en bloc to the 'evasions of bottom fact', as a Standard Oil executive once famously advocated to Rockefeller in a memo? Will it deploy the bribes masquerading as political contributions, the secret agents, the mercenary coalitions with foreign governments?

These were the questions I mulled over with Paul Hohnen, as we wandered in the killing fields of Virginia. We pondered on what we had seen. All of it so far rather pointed to depressing answers.

Back at the conference centre, the evening's reception was hosted by the Global Climate Coalition. Well over a dozen industry representatives were there, plying the delegates with good food and wine. Bold namecards showed which company they represented, oil and coal companies for the most part. Paul Hohnen and I circulated, determined to eat as much of their food as we could. I stood next to a group where a man from the World Coal Institute was talking in a loud voice to the Samoan Ambassador about the IPCC report. 'It is careless science, you know,' I heard him say.

The days ticked excruciatingly by. The climate treaty – to be called the Framework Convention on Climate Change – was due to be finished, and ready for signature by heads of state, by the time of the Earth Summit in June 1992. Then, world leaders were due to meet at Rio de Janeiro for the biggest and most important gathering ever to discuss environment and development. This target now began to look unlikely. Only four two-week sessions of talks were scheduled between now and then. What the governments of the world had achieved – after two weeks and an estimated cost of $2 million – was merely a decision to arrange future negotiations into two working groups, one on commitments and related matters, and the other on implementation. And the chairs of the working groups had not even been elected.

Meanwhile, in the desert sands above 70 per cent of the world's proven oil reserves, world affairs were moving somewhat more swiftly.

JUNE 1991, ROVENIEMI, FINLAND

Detecting an undeniable fingerprint from humankind's greenhouse-gas emissions was going to be problematic for the IPCC's scientists because of the natural variability of the global climate. The warmest years in over a century of records had clustered in the 1980s, but in 1990 most climatologists insisted that this could still, in principle, be mere coincidence. In their 1990 report, the IPCC concluded that it would be more than a decade, in fact, before a clear signal would be seen above the noise of potential natural variability. But it was anomalous weather – the drought in the US Midwest – that had contributed to the creation of the IPCC in 1988, and quite a catalogue of suspicious warming phenomena was now appearing around the world.

California had gone five winters without serious rain, suffering the worst drought in the state for a hundred years. Private reservoirs had run dry, stringent water rationing was in force, and voters placed water second only to crime when rating their concerns. Economic losses amounted to $3 billion and counting. In Antarctica, the Wordie

Ice Shelf had begun to disintegrate. Atmospheric warming had freed huge icebergs into the Southern Ocean. Spectacular satellite images of these drifting monsters appeared in *Nature* magazine during March 1991, with a warning from British Antarctic Survey scientists that if the warming trend continued, other ice shelves along the Antarctic peninsula might be at risk.

In April, a paper in *Nature* by an international team of scientists reported water temperature measurements in the Soviet Arctic in 1987 and 1990. The water was fully a degree Celsius warmer in 1990 than it had been three years before. Was this a signal that the Arctic was warming, or was it just variability from year to year? The scientists weren't guessing. But one thing was obvious to them: it seemed clear that there was a need for careful monitoring of heat fluctuations in the Arctic.

Given my fears about warming shallow Arctic waters being able in the future to melt methane hydrates, this report filled me with particular concern. In Arctic methane, we were dealing with one of the biggest potential long-term positive feedbacks in a warming world.

The nations bordering the Arctic were to meet in June 1991 in the small Finnish town of Roveniemi to sign an agreement called the Arctic Regional Protection Strategy. The USA, Canada, the USSR and the Scandinavian countries all sent ministers to Finland to complete the negotiation of the text, and to sign the treaty. I was there too. My mission was to see what could be done to persuade the American and Soviet governments to have their navies offer up their water temperature and ice thickness data for public scrutiny and use by climate researchers.

Among swarming mosquitoes the size of butterflies, I enjoyed remarkable access to ministers and diplomats in Finland. The Soviet environment minister at the time was a biologist called Vorontsov. I managed to catch him alone in a corridor, and asked him straight out why the Soviet navy couldn't embarrass the Americans by making public the data from their redundant missile submarines. This, after all, was the new era of glasnost.

He told me that as a scientist he was as worried as I was about what might be happening in the Arctic. He had already tried persuading the Defence Ministry that the data were needed by climate modellers, among others. He had got nowhere.

'Couldn't you ask Gorbachev to intervene personally?'

Vorontsov snorted. 'You really think he has any influence over these people?'

It wasn't just the cold zones that were throwing up worries about warmer than normal seawater. Corals are very temperature-sensitive organisms: in most oceans they thrive in waters of around 27–28°C, but are singularly unhappy when the waters become just a degree or so warmer. Heat-stressed corals expel the algae that normally live in their bodies, and from which they obtain food; they turn white as a consequence, in a process known as coral bleaching. If the stress is maintained for long, they die. During the 1980s and early 1990s, every major reef system had been affected by coral bleaching at one time or another. A number of other stresses can cause the bleaching effect: sewage pollution, exposure to ultraviolet light, even water colder than normal. But many of the recent episodes had demonstrably been associated with waters warmer than they should normally have been. After an unprecedented episode of coral bleaching in the Caribbean in 1989, four US reef experts had testified in Congress that, in their opinion, corals were acting as canaries-in-the-coal-mine, providing the first signal of global warming.

The phenomenon could assume the dimensions of ecological catastrophe. In 1990, corals in the southern part of the Arabian Gulf had suffered 50 per cent mortality. In particularly warm waters around the Galapagos Islands in 1983, over 90 per cent of the coral died in some areas.

As usual, there were potential extenuating circumstances to cloud the interpretation of what coral bleaching and unusually warm tropical water masses might mean. In the Pacific, the semi-regular meteorological phenomenon El Niño, which generally takes place every three to five years or thereabouts, is marked by unusually warm waters. During an El Niño event, for reasons which are not at all clear to scientists, a warm body of water builds up in the eastern Pacific over a period of a year or more. This causes a weakening or reversal of easterly trade winds, and a turnaround of normal climatic conditions: in particular, the west coast of South America experiences heavy rains, and Australasia suffers drought. The 1982–83 El Niño event, for example, was the strongest of the century: it brought among other things the worst-ever droughts in Australia since the 1860s, and the mass die-off of Galapagos coral. El Niño events were happening well before humankind began enhancing the greenhouse effect, so there is no doubt that they are natural, even if lacking an obvious explanation. However, by 1991 meteorologists were beginning to voice fears that global warming might mean stronger El Niños.

An El Niño had begun in 1990. Now, in April 1991, reports came out of French Polynesia that corals were suffering widespread bleaching. For several weeks, fluorescent colours had been spreading like a rash across the reefs around Tahiti. The ailing corals were erupting carpets of brilliant whiteness. The average water temperature at the time of year was normally around 28.4°C. Now, it was 29°C. Local scientists, according to French newspaper reports from Tahiti, were estimating 85 per cent mortality in some areas, and blaming global warming.

The reports came to me via Washington, where American environmentalists were trying to use this tragedy in tropical paradise to turn the Bush administration from its obstructionist course at the climate talks. My depression at what I read was tinged with a reluctant hope that Bush's White House would now be forced to take the global warming issue a little more seriously. This was, after all, tantamount to a mass epidemic in the second most diverse ecosystem on the planet, after the rainforests.

JUNE 1991, GENEVA

At the second session of climate negotiations, the first order of business for governments was to clear up the unfinished matter of who would chair the two working groups. The negotiations broke up into a series of closed consultation meetings to search for a deal that served the interests of as many countries as possible.

The hours turned into a day. And then another. After three days, the German delegation issued a memo complaining at the slow progress.

'Further delays,' said the head of delegation, 'would be irresponsible.'

Deals between the industrialized countries and the developed countries, and between the different UN regions, were eventually struck. It had taken three weeks to decide even on the shape of the negotiating table.

On 23 June 1991, less than a year away from the Earth Summit in Rio where the final Climate Change Convention was supposed to be signed, talks finally began on the treaty itself. The first attempt to identify a route to consensus came from the Japanese delegation. They called their new idea 'pledge-and-review'. It aimed to try and bridge the gap between the White House, with its 'Just say no' approach, and the rest of the industrialized world, which sought legally binding commitments on emissions, with specific targets and timetables. Under pledge-and-review, states would sign a convention devoid of any commitments at the Earth Summit. They would pledge what they could in the way of targets, and agree to review their commitments, and the implementation of those commitments, at an interval to be agreed.

At the end of the session, the industrialized nations agreed to take this compromise proposal away and analyse it in time for the next session. Environmentalists left Geneva fearing that the intransigence of the Bush Administration was already setting the talks on course for a toothless compromise treaty at best.

JUNE 1991, LONDON

I flew back for my most stressful assignment yet. The Royal Institution for International Affairs had invited me to debate with the leading global warming sceptic, meteorologist Professor Richard Lindzen of the Massachusetts Institute of Technology. An invited audience of senior industrialists would be there.

Scientists sceptical of global warming had become relatively well known around the world by this time. Regular media performers included Fred Singer and Patrick Michaels of the University of Virginia, Robert Balling of Arizona State University and Sherwood Idso of the US Water Conservation Laboratory. These sceptics were in a clear and tiny minority. None the less, wherever you saw an IPCC scientist being interviewed on TV, you would very often see one of the doubters lined up opposite.

Richard Lindzen stood head and shoulders above the other sceptics in terms of intellectual weight. A world-class meteorologist, Lindzen's central idea was that water vapour would be wrung out of the upper troposphere by the increased vigour of atmospheric circulation in a warming world – a negative feedback which would make all the other feedbacks look insignificant. His idea had been considered by the IPCC, which had looked at satellite data from a programme called the Earth Radiation Budget Experiment, and other evidence which pointed exactly the opposite way – to a strong positive feedback from water vapour – and given him the thumbs-down.

Lindzen was a suave and fluent talker, and a man with a very convincing manner. He tried throughout our debate to draw me into his domain, the dynamics of water vapour in the atmosphere, where he knew I had no experience. For my part, I tried to draw him out into the holistic view of climate – to consider potential feedbacks of which he himself had little experience.

I asked Lindzen what I see as the critical question for those who doubt the IPCC's basic conclusions. Given that he did not question the heat-trapping ability of the greenhouse gases, or their build-up in the atmosphere, how sure was he of his own theoretical suppression mechanism for global warming? Was he 100 per cent certain? Because if he was, I wanted to know how it was possible. I was under the impression that no scientist could be absolutely certain of the specifics of how an artificially enhanced greenhouse effect would play out. Was he 80 per cent certain? That would be a very high degree of confidence. Yet, if so, in maintaining there was no need to cut greenhouse-gas emissions, he was still asking us to take a one-in-five chance that he was wrong. And if he was wrong about his own favoured feedback, there were plenty of potentially huge positive feedbacks waiting to tip the world towards climatic chaos, and eventually, perhaps, intolerable heat. Given the ultimate stakes, that was simply not good enough.

We debated back and forth on this theme. Finally, Lindzen conceded a key point: 'I suppose it's possible there will be unprecedented climate change,' he told the Royal Institution's audience, his face screwed up as though to convey the vanishingly small odds he attached to this prospect. But he would not be drawn into answering the question of how confident he was of being right, or into an examination of the implications of that. While there can be no doubting the sincerity with which he held his views, Lindzen, I came to understand, saw the whole debate as a matter of simple academic dispute. Of course, scientists like him, out on a limb in the face of a peer group which holds a robust counter-view, have ultimately been proved correct on many occasions during the history of science. One can think of Galileo, and even Einstein. But the stakes are a little different with global warming. The future of civilization, perhaps even of life on Earth, did not potentially hinge on Galileo or Einstein being right.

Meanwhile, Richard Lindzen continues to accept all-expenses-paid trips to fly around the world and speak wherever the carbon club thinks he needs to be listened to. Less than a year after my debate with him in London, he appeared in Vienna to speak at OPEC's first-ever conference on the environment. OPEC didn't invite anyone to present the opposing view.

JULY 1991, TOKYO

In my old life as an academic, I had once seen the Yangtze river. It was sometime in the early 1980s, and I was in Nanjing, on a British Council lecture tour of China. I shall never forget my first sight of that river: more vast than I could have imagined possible so far inland, a thousand-lane highway of surging yellow eddies, as much liquid mud as water. My companion, from the faculty of earth sciences at Nanjing University, had taken me to the mid-point of the astonishing bridge spanning the river at its lowest bridging point. 'Made in China,' he told me softly as I looked in wonder north and south to the far-off banks. He told me exactly how many millions of rivets and tonnes of steel it had taken to build the bridge. I asked him how he knew so much about the bridge – he was an earth scientist, not an engineer. He had helped build it, of course, he and his faculty colleagues, the professors among the labourers, during the Cultural Revolution. Several of his friends had been among the many who had died in the process.

In July 1991, the Yangtze claimed another thousand Chinese lives. The worst floods of the century were bursting across the Yangtze plain. An incredible 20 per cent of China's croplands was in the process of being submerged. Ten million Chinese were being made homeless, and more than half the billion-plus population of China would end up being adversely affected in some way. Chinese scientists, according to agency press reports, were blaming global warming.

A country which faced the challenge of feeding a quarter of the world's population with 8 per cent of the world's cultivable land had lost 20 per cent of its croplands to one extended period of unusual rainfall.

As the floods spread, I flew to Tokyo. My Japanese Greenpeace colleague, Yasuko Matsumoto, had arranged for me to speak in the Japanese parliament, at a meeting of an all-party international group of politicians called Global Legislators for a Balanced Environment. GLOBE, surely one of the world's more dreadful acronyms, had presented me with an opportunity the full extent of which I could not know at the time. The President of the GLOBE organization was a certain Senator Al Gore.

Attending this annual meeting of Gore's group were several dozen American, European, Japanese and Russian politicians, of all political persuasions: Republicans and Democrats, Christian Democrats and Socialists, Liberal Democrats and Communists. They were unified to a surprising degree, I was to find, by their concern about the environment. Al Gore began the meeting by giving a long speech. His GLOBE colleagues sat listening to him in the chamber of a meeting room in the parliament. Observers and journalists packed the public galleries. This was to be the first of several occasions when I was to see clearly, at first hand, the future US vice-president's knowledge of, and concern about, global warming. I was more than impressed.

Soon afterwards, I delivered my own speech. The achievement of my colleague Yasuko in setting up such an opportunity cannot possibly be understated. Shortly before she joined Greenpeace in 1989, the organization had been described by a Japanese government minister as a 'terrorist group' on account of our harrying of a Japanese whaling fleet or two with inflatable boats on the high seas. By 1991, however, Yasuko had corrected such misapprehensions, at least in the Environment Agency. The environment minister operated a virtual open-door policy towards her, and it was her reputation, in government and media circles, that had presented me with the opportunity of addressing GLOBE.

I built that speech in the Japanese parliament around insurance against a worst-case analysis for global warming in which all the feedbacks ended up being positive, and acted in synergy. Such a scenario might begin, I suggested, as the oceans gradually warm, with a slowing down of circulation. The planet would then be suffering a steady undercutting of its ability to store heat in deep water. At the same time, phytoplankton volumes would be declining in nutrient-starved surface waters, decreasing the ocean's ability to draw carbon dioxide down from the atmosphere. Meanwhile, perhaps, the changing balance between photosynthesis and respiration in land plants might add to the problem. Warmer temperatures boost both respiration and photosynthesis, but the former more so than the latter. In a warming world we can reasonably expect more carbon dioxide to be produced than is sequestered, causing further net accumulation of carbon dioxide in the atmosphere, further boosting temperatures. On top of this, as forests become drier, forest fires might break out more frequently. Even where there are no severe wildfires, forests might come under attack in some areas as a result of increasingly ferocious windstorms.

Then, suppose the tundra began to melt. The wetlands would spread in Siberia, Finland, northern Norway, Canada and Alaska, and in them accelerating oxidation of organic matter might cause huge amounts of carbon dioxide, and methane – a more potent greenhouse gas than carbon dioxide, molecule for molecule – to be driven into the atmosphere.

Cloud feedbacks are a major source of uncertainty in climate models. Suppose that here our luck is out as well. The thin, high-altitude clouds of the warming world might act to trap more heat than the reflective lower-altitude clouds can cast back into space from incoming solar radiation. In the Arctic, the ice cap would at some stage start to melt, also cutting down the amount of solar radiation the planet reflects, so providing more heat to be trapped.

By this stage, the stressed planet would be emitting more and more greenhouse gas of its own, reducing the effectiveness of any particular human policy aimed at slowing the build-up of greenhouse-gas concentrations in the atmosphere. By now, there would be profound aridity at many latitudes. We might then risk losing the 'fertilization effect'. Some plants grown in a glasshouse with excess carbon dioxide in the air grow more robustly, storing more carbon. Such a fertilization effect could well be in operation in growing forests today, coming to our rescue somewhat and slowing the build-up of carbon dioxide in the atmosphere. We do not know for sure: the natural world is very different from a glasshouse experiment. Suppose we discover that for many years excess amounts of carbon dioxide have indeed been taken out of the air by growing forests. Such a mechanism for suppressing global warming would be a finite process – it couldn't go on for ever. As some trees reach the limit to which they can store carbon, while others succumb to the spread of drying soils, the fertilization effect could become a thing of the past.

Yet still the bad luck might continue. The chemistry of the troposphere could begin to change to our disadvantage. It can seem to people as though all the minor constituents of the atmosphere are bad for the planet: trapping heat, or depleting ozone, or causing disease when breathed in polluted cities. But this is not so. One minor constituent, called the hydroxyl radical, is hard at work, as fast as humans pollute the atmosphere, trying to clean it up. The hydroxyl radical oxidizes – and so neutralizes – gases like methane, carbon monoxide, nitrogen oxides, HFCs (hydrofluorocarbons) and HCFCs (hydrochlorofluorocarbons). It is the atmosphere's 'cleansing agent'. The trouble is, there isn't much hydroxyl to go round, and like so much else it is poorly understood: in this case because of the complex chemistry of so many interacting gases. There are many ways in which the hydroxyl radical can be used up, but there are also many in which it can be re-formed. The fear is, of course, that it will begin to be less effective as more pollutants build up. If the hydroxyl radical does begin to be overwhelmed, then we have a serious problem. The atmospheric lifetime of greenhouse gases – the period of time they stay up in the air, trapping heat – will extend. Global warming will accelerate.

If such feedbacks did indeed gang up on us, then at some time in the future a bad-dream world would become a nightmare. Suppose governments then moved from banning cars in cities, through banning Sunday driving, to completely outlawing all but solar-powered battery, fuel cell, or other 'zero emission' cars. Suppose they decided to build no more coal-fired power stations, only solar and other renewable-energy ones. Still the treadmill out there in nature might not stop. And then a far worse feedback might begin to stir. Below the tundra and on the shallow floor of the Arctic, in sediments at and just below the seabed, are peculiar substances that many people have never heard of. They are called methane hydrates.

A methane hydrate is a solid, ice-like mixture of water and methane, which forms when temperatures are low and pressures sufficiently high. The methane molecules are literally trapped, under pressure, in a cage of water molecules. A hydrate resembles ice, but isn't ice: the water crystallizes differently, and hydrates can form above freezing point if the pressure is sufficiently high. Hydrates will form, for example, if gas is pumped under pressure through pipelines in cold conditions, clogging them up. They will also form under sediment or seawater and, in cold enough conditions, will do so at fairly shallow depths. Methane generated from organic matter, and leaking from natural gas reservoirs, can be trapped in and below hydrates over thousands of years. But if the temperature in the surrounding water or sediment rises to the point where a methane hydrate becomes unstable, methane gas is released at once.

Methane hydrates show up on seismic profiles – pictures of structure below the seabed made by bouncing sound waves down from ships. I have seen them, like thousands of other earth scientists who have worked on the history of the oceans, almost wherever there is deep water to provide the necessary pressure, and sediments thick enough to cause the natural accumulation of methane. I have even seen a natural methane hydrate face to face. I must be one of a relatively small number of scientists to have done so. In 1979, I was on an American drilling ship which pulled methane hydrate up in a core off Mexico. Transported from the domain of high pressure and low temperature where it was stable, more than a kilometre below the sea's surface, it literally vanished before our eyes – a frothy cylinder of dirty ice, fizzing as it released the methane trapped under pressure within. The drilling engineer told us we had to stop drilling at once. He was not allowed to drill on, for fear that we would go through the solid methane hydrate layer, release the gas trapped below, and threaten the ability of the ship to float.

How much hydrate is there, then, around the planet? And how much of it might be in danger of melting in a warming world? The US Geological Survey's estimate for the total amount of carbon in the methane hydrates of the planet is 10,000 billion tonnes, making it one of the biggest reservoirs in the carbon cycle. Most of this, thankfully, is out of the reach of global warming, in sediments under deep ocean where temperatures are unlikely to change much, no matter what happens at the surface. But where the water is particularly cold, less pressure is required for hydrate to form. In the Arctic, hydrate can be found below water as shallow as a few hundred metres. The question is, how much hydrate lies around the Arctic? We do not know for sure, but it must be measured in many tens if not hundreds of billions of tonnes. And since there is a mere five billion tonnes of carbon in today's stock of atmospheric methane, only a little methane hydrate would need to be melted to boost the greenhouse effect significantly.

Suppose the seas around the melting ice cap did indeed become warm enough to do this. Around the world, monitoring instruments would show a dramatic rise in atmospheric methane, alongside the carbon dioxide. The UN would meet in emergency session, perhaps. Every possible method of cutting human greenhouse-gas emissions might then be made mandatory. Things that would seem laughable today, like jail sentences for burning coal and oil, might become commonplace. But it might be too late. No matter how deep the cuts in human emissions, they might be cancelled out by the natural emissions of carbon dioxide and methane we had enforced by artificially raising the planet's thermostat.

I explained to Senator Gore and the parliamentarians of GLOBE that it didn't have to be like that, of course. We were talking about the very worst case. The point was not that it would happen, only that it could happen. Policymakers knew all about insuring against worst-case analyses to make sure they didn't happen. Take the worst-case analysis of an invasion of Western Europe by the Red Army. Yet how many officials at the climate negotiations understood the worst-case analysis in assessing the threat of global warming? How many knew what a runaway greenhouse effect was, much less that it was possible? All the Intergovernmental Panel on Climate Change had talked about was 'the scope for surprises'.

I could see as I wound up that a number of the parliamentarians were itching to ask questions. One, US Congressman Jerry Sikorski, shot his arm into the air as soon as I stopped. As well as listening, he said, he had been reading my account of a worst-case scenario in the paper I had written for the occasion. He urged his colleagues to read it. It was, he said, like a Stephen King novel, only with a difference. It wasn't necessarily fiction.

JULY 1991, CANBERRA

From Japan I flew to Australia, where the agenda included some hard lobbying of the number one coal exporting government. Canberra still had a national target for reducing carbon dioxide emissions, but only just, and Paul Hohnen and I had to work the ministries to do our best to shore things up.

We went armed with a copy of an 8 July *Time* magazine report on the state of the Lloyd's of London insurance market, the world's only physical market for insurance, where brokers meet underwriters face to face to place business. 'Lloyd's is reeling,' said the headline, 'and as the fine print catches up with them, many investors face financial ruin – down to the last cuff-link.' The *Time* article showed clearly how catastrophes involving extreme climatic events were a major part of the problem emerging in the three-hundred-year-old institution. Between 1967 and 1987, the insurance industry had sailed along without facing a single natural catastrophe costing them a billion US dollars, anywhere in the world. Then such 'billion dollar cats', as insurers call them, had started turning up in a rush: the October 1987 windstorm in north-west Europe, with a $2.5 billion bill; Hurricane

Hugo in September 1989, at $5.8 billion the most costly windstorm ever; the January and February 1990 European windstorms, with total losses exceeding $10 billion; then the July 1990 storms in Colorado clocking up $1 billion. For Lloyd's, the impact of this catalogue of misfortunes had been severe. The long period without 'big cats' had had the effect of attracting capital and competition, and reducing the price of reinsurance. Capital flowing into Lloyd's had been around $2 billion in 1978, but by 1987 it exceeded $10 billion. Much of this came from rich investors who saw the market as a route to super riches, and signed up for unlimited liability.

But in 1988, Lloyd's moved from billion-dollar profits to billion-dollar losses. In 1989, losses reached $3.3 billion. The 1990 loss would not be announced until 1993 because Lloyd's accounting requires three years for all claims to be completed, but it was already known to be huge: hence the article in *Time*. It would later turn out to be $4.3 billion. Investors in the biggest-losing syndicates were facing total ruin, and their lawsuits, alleging negligence, were hovering over the biggest-losing underwriters. Several investors had committed suicide.

All this, Paul Hohnen and I told the havering defenders of status-quo coal use in Australia's civil service, could easily end up being just a microcosm – a foretaste of the dawn of the beginning – of the economic problems that would come with global warming. Unless, that is, the nations of the world could together do something about their reliance on coal and oil.

AUGUST 1991, FEDERATED STATES OF MICRONESIA

From Australia, I flew to the annual South Pacific Forum summit. At the end of the meeting, the Pacific heads of state once again issued a plea for action on climate change from the industrialized countries. Australia and New Zealand put their names to it, as they had in 1990: but this year with a clear reluctance.

I took off for a few days after the summit, to try to recharge some by now very flat batteries. I found an ecotourism hotel picking its way via a series of bamboo walkways through the jungle on a volcanic hillside overlooking a spectacular coral-spangled bay. I sat on the terrace of that hotel for hours, and simply looked out across the water. As the evenings fell, so the most spectacular sunsets I have ever seen would spread in violets, reds, oranges and yellows across the sky. The beauty of the scene from the terrace, fringed in palm fronds, and with a soundtrack improvised by parrots, was uncapturable by word, much less by photograph.

Yet even this I could not enjoy without a mental filter. The sunsets were this brilliant because of the Mount Pinatubo eruption in the Philippines in June, which had thrown millions of tonnes of dust and aerosols into the stratosphere. The eruption was the largest of the century. The world had probably not seen anything bigger since Krakatoa blew its top in 1883. The cooling effect of big volcanic eruptions being well known, there was little doubt that the Pinatubo ejecta which were then playing kaleidoscopic games with the sun's evening rays would cool global temperatures for several years, so suppressing the emerging concern about global warming. The eruption of Tambora in Indonesia in 1815, which injected ten or fifteen times more sulphur into the atmosphere than Pinatubo, led to a terrible summer in the northern hemisphere, with frost in New England. The eruption of Toba in Sumatra 73,000 years ago was even larger, and may have accelerated the onset of the last ice age.

I didn't know it on my idyllic terrace in Micronesia, but at the time NASA's top climate modeller, Jim Hansen, was predicting that the Pinatubo dust would cool the planet by fully 0.5°C. By the end of 1992, he would be shown to be bang on target. Satellite images were to show the aerosols hanging around in the stratosphere as a shimmering band round the equator, busily reflecting solar radiation back into space and so turning down the planet's thermostat. They would continue doing so until late 1993. Throughout that time, of course, they would be masking the heat-trapping effects of the greenhouse gases, dampening global temperatures, and delaying the day when society would awaken fully to the danger it faced.

SEPTEMBER 1991, NAIROBI

I flew to Nairobi for the third session of climate negotiations, staring down from a jumbo jet at the barren wastes of the Sudan, like a map in shades of ochre below me, thinking about the impact of global warming on yet another part of the planet I was setting eyes on for the first time in my life. The climate diplomats gathered in the barn-like negotiating hall in the headquarters of the United Nations Environment Programme, just outside the city. If the Convention on Climate Change was to be signed as intended at the Earth Summit, now less than 10 months and just three negotiating sessions away, things were going to have to start speeding up soon.

The Japanese had come armed with a document which put flesh on their pledge-and-review concept. The wording was typical of the contortions governments were getting themselves into in trying to find a position which bridged the gap between the USA and the rest of the industrialized world. The delegations carried it off into a hundred separate meetings to decide how to react, and on the third day the European Community rejected it outright. The Europeans were holding their ground, and now made a key move. They proposed that the objective of the convention should be to stabilize atmospheric greenhouse-gas concentrations at levels that would not dangerously interfere with climate. Specifically, ecosystems should be able to adapt, food supplies should not be threatened, and economies should be able to prosper. This was an entirely obvious, but environmentally tough, goal to set.

Paul Hohnen and I were now presented with an opportunity to stir the disharmony between the EC and the USA. We had been faxed a copy of a letter sent by Senator Al Gore to the US head of delegation, Bob Reinstein, some time earlier. In it, Gore complained in strong terms that Reinstein had 'seriously misrepresented' European opinion during recent testimony to the US Senate. Commenting on the previous session of climate negotiations in Geneva, Reinstein had evidently testified that EC countries had relaxed their insistence on carbon dioxide targets and timetables. Furthermore, he alleged, many had expressed support for the Japanese pledge-and-review proposal. This, we knew, was guaranteed to arouse more than a few European indignations. Paul and I rushed to a photocopier. In no time, stacks of the copied letter were being vacuumed up by delegates, and frowns could be seen creeping across European faces.

Each day at the negotiations, the environmental groups produced a newspaper called *Eco*. Copies of this publication were piled up at the back of the negotiating hall each morning, and would soon be depleted by delegates, who knew that the contents of *Eco* were also committed to cyberspace, not to mention the fax machines of journalists all over the world. In Nairobi, the gossip column in the newspaper was given the name 'Tusker'. The day after the leak of Gore's letter, the Tusker headline read 'REINSTEIN GORED'.

Bob Reinstein and Dan Reifsnyder of the US State Department reacted interestingly, for diplomats representing their country on matters of life and death. They wore sweatshirts to the negotiations, designed by the US Department of Energy for another purpose, but evidently deemed applicable here. They showed a frowning American Eagle holding up a finger-like feather with one wing, and clutching a screw in its other wing. The eagle glared out from the crest of the Department of Energy. The caption read, 'Trust us, would we lie to you?'

Reinstein wrote back to Gore. He had never intended to reflect a view that the EC had changed its substantive goal, he said, merely that the Europeans had not been seen to press hard on carbon dioxide targets and timetables. Armed with a pile of copies of this missive, one of Reinstein's enthusiastic underlings in the State Department approached Paul Hohnen. 'You seem to be good at distributing things,' she said, looking at a point in space around his left armpit. 'Why not distribute these?'

On Wednesday 12 September, the United States gave its long-awaited statement to the negotiations on commitments. Bob Reinstein didn't speak much in the sessions. He could afford to let Saudi Arabia and Kuwait be seen to do the dirty work. The USA, he announced, had carefully examined the prospect of stabilizing or slightly reducing emissions of carbon dioxide. 'Targets and timetables for reductions of specific greenhouse gases are one kind of signal,' he concluded, 'but by no means the only kind of signal, and, in our view, not necessarily the most appropriate signal at this stage.'

*

On 27 September 1991, the eighth billion-dollar windstorm in four years hit the insurance industry. Typhoon Mireille was the sixth strongest in the Japan Meteorological Agency's records, and no typhoon as strong had hit Japan for 30 years. It damaged 1.6 per cent of all Japanese households, draining $2.21 billion from the Japanese insurance industry's property-catastrophe reserve system, leaving just $1.86 billion at the end of fiscal 1991. Globally, insured losses from this one event reached $4.8 billion. The storm skirted the western edge of the Japanese archipelago. In other words, it could easily have been much worse.

The cyclone had two particularly interesting meteorological characteristics. While crossing the Philippine Sea, it was forced to move to the north-west around a strong area of high pressure east of Japan, so that it swung to the east, as such windstorms are prone to do in that region, further north than usual. Second, it did not lose strength once it made landfall in south-west Japan, and this, Japanese insurance-industry analysts would later report, was probably because the eastern China Sea was fully a degree Celsius warmer than it normally was for the time of year.

NOVEMBER 1991, LONDON

Shell, the largest oil company in the world in 1991, takes great pride in its ability to plan ahead. Shell's planning department employs people of formidable intellect merely to sit around and think about the future. They call their business scenario planning. One day in 1990, some of Shell's scenario planners invited me for a business lunch. Three men who could spout energy statistics as though they had an invisible atlas in front of them grilled me for more than two hours. Eventually, they tried to sum up our meeting. 'We agree with you that Shell will not be an oil company for ever,' the most senior of them offered, encouragingly. 'The future will be in alternative supply and energy services. I think the difference between us is the timescale. We think we have more time than you evidently do.'

Like a second oil century, more or less.

In November 1991 I found myself once again in Shell's headquarters.

A long-lapsed friend, a fellow graduate student in 1978, had arranged this particular visit. He now found himself a buyer and seller of Shell's oilfields, and organizer of their seminar series. I met my old colleague in the foyer of the great grey edifice on the Strand which is Shell House. Save for the executive pinstripe, he had not changed since I had seen him last, maybe ten years before. He had managed oil rigs in Brunei, he told me. He had explored for oil in the North Sea. Now, he analysed risk. Risk in the buying and selling of oilfields, that is.

He showed me to a room labelled 'cinema'. There was to be no coffee or socializing, evidently. We chatted uneasily about former contemporaries at Shell International and Shell UK as people filed in to fill the hundred or so comfortable seats.

'You've got some big names,' said my host, interrupting himself. The exploration manager had walked in.

He introduced me to a full house. I delivered my stump speech. A few days before, I had given the same presentation to an audience of eight hundred oil industry personnel at a Shell-organized conference in The Hague. I had been asked on both occasions to leave a lot of time for discussion.

The first question was on biomass. Couldn't fossil-fuel carbon dioxide be soaked up by massive programmes of afforestation? I gave the standard answer. Growing and regrowing forests could make an important contribution, but only a marginal one. This was no magic bullet, no substitute for the deep cuts that will be needed in fossil-fuel burning.

Then came the standard critique of renewables, and how they weren't up to the job. There was much we could do with existing renewable-energy technology, I countered, never mind the kind we could develop if we spent as much on renewables R&D as we do on fossil-fuel and nuclear R&D. We were at the crisis-management stage: we needed a Manhattan Project on renewables and energy efficiency.

I saw one man in the audience nod his head. I knew from my experience in The Hague that there were many in the industry quite prepared to agree with this view in private.

As I had experienced so often when talking to oil industry audiences, there was a lot of blame-switching. The government had to act first, said one. The public had to change its behaviour, said another. Anything but proactive action by the oil industry.

Then came an American voice. 'I don't like to hear the US picked on as much as you have in this presentation. It's not true that Americans drive low-mileage cars for preference. The Japanese sell millions of low-mileage cars.'

From five rows back, an English voice interrupted him. 'That's simply not true.' The American turned to him, a frown on his face. And they began to argue among themselves. I watched in astonishment. The American became long-winded in defending his turf. The audience visibly loosened up, enjoying the developing argument. Another chipped in. Then another, this one exercising wit, playing for laughs. And as the minutes clicked up, so did my amazement. I had come here expecting a grilling, and instead I had instigated an in-house debate.

The next question was on my contention that Shell should extend its corporate planning horizon: the now much-ventilated tactic of asking them to explain how they were going to be doing effective business in the globally warmed world of 2030. That would very probably be a world with economies beginning to implode under the multiple environmental stresses they would be facing. The corporate planning horizon was only five years, so the Corporate Planning Department had told me. If they extended that, I proposed, they would find that facing the challenge of a paradigm shift in the way they did business would look increasingly attractive. And if they positioned ahead of their competitors, they would profit enormously in newly emerging sectors of the energy business.

'I don't think you are entirely fair to say that we plan at most on a five-year time frame,' a man near the front told me. 'In reality, it's longer than that.'

Again there was an interjection from the back rows. 'Oh, come, come. Let's be honest. We say it's five years, but the truth is we're lucky if it's even one.'

And most of them laughed.

In the Geological Society's imposing headquarters in Piccadilly, the geological community holds its conferences, hands out its accolades, and steeps its new recruits in the traditions of the profession. How I had loved this place, the Mecca of geology for academics and industrialists alike, when I had been one of them. Under its portraits of Darwin, Huxley and the other giants of the past I had presented my papers, the edge of academic ambition no doubt clear in my voice. From its plush seats I had risen, twice, at the annual President's Evening to receive awards for my research, in front of the cream of my peer group and two proud parents. But on this day, some years on, the job allocated me by the Geological Society was to address a careers fair on the ethical problems facing the oil industry.

The audience was to be two hundred soon-to-graduate trainee geologists, and there was to be one other speaker: the chief geologist of Amerada Hess, a company whose logo, by dint of large donation, appeared on a plaque in the entrance to the Society. His would be the story of the glittering prize: the career at the front end of a booming industry. Mine would be the story of that career in a broader context. A nag at the conscience, a lacklustre and unappealing story, no doubt, to the student with an overdraft at the bank and ambition in the heart.

The oilman would speak before lunch, and I would speak immediately after. There was, the executive secretary of the Geological Society told me, no need for me to attend the morning session. They all knew how busy I must be.

That morning, I was researching at the Institute of Petroleum. In the industry's in-house literature there was much of interest to the environmentalist. I had a stiff deadline for a report, but the going was proving hard. I felt the need for an early cappuccino. I decided that I might as well find one en route to the Geological Society, and go on to hear what the opposition had to say.

I crept into the back of the darkened lecture theatre soon after the oilman's lecture had started. The speaker looked very young for a chief geologist: younger than me, and none of my student contemporaries had yet made chief geologist. He looked vaguely familiar. The auditorium was full, and so was much of the standing room. I found a place where I could lean against the back wall.

The speaker was analysing the energy future. Come what may, he said, the global demand for energy would rise by at least 1 per cent per year. How, he asked, could that demand be met? There followed a tour of the various options for renewable energy. He began with wind power, and showed a splendidly coloured, perfectly designed slide illustrating how vast an area would need to be covered with wind turbines in order to generate the electricity requirements of one good-sized modern city. No mention of the energy wastage in a modern city, or the fact that there was nothing to stop farmers making hay, literally and metaphorically, around the wind turbines. No mention, either, that generating electricity is not a game in which oil has much relevance.

Besides, he added, windmills are noisy, unpleasant, and potentially lethal . . . especially if you are a seabird.

The students laughed.

As for solar thermal and solar photovoltaics, if we relied on solar power, what would we do in the dark?

More laughter from the audience, and some were even scribbling down these words of wisdom.

No mention of batteries, hydrogen, fuel cells, a mix of renewables – even hybrid systems. No mention that solar photovoltaic energy depends on light, not sunlight. Solar PV operates less effectively on cloudy days, but it still works.

And so the assault of selected anecdotes continued. Tidal power, for example, would be devastating for seabirds. *Exxon Valdez* wasn't?

Even nuclear did not escape. He had played golf in Kent the other day, with a nuclear power station just across the water. That, he said, was something that frightened the life out of him.

These are the alternatives, he said, summing up. The words 'energy efficiency' or 'demand management' had not passed his lips. Oil had a glittering future, then, and the good news was that there was more of it around than had been calculated a few years ago. 'The dreaded day that oil runs out has been pushed into the future. You can enjoy a full career in the oil industry. We are entering the innovative phase. The more innovative you are, the longer that phase will last.'

Of course, we had to consider the environment, he said. Cue global warming. He showed only one slide, a plot of inferred sunspot activity alongside a plot of global average temperatures in the twentieth century. He had taken it from a paper published shortly before by two Danish scientists in *Science* magazine. There was no mention of the IPCC, which had concluded that variations in solar activity can constitute only a minor forcing of the Earth's radiation balance compared with the heat-trapping capacity of the greenhouse gases. The inference was clear: the global warming argument was an annoying diversion, one which would disappear in good time as more reliable science, like this Danish study of sunspots, became available. He did not offer a definitive condemnation of global warming (they rarely do). 'As responsible scientists, of course, we can't afford to take any chances.' But the pitch for caution ran on to become a plug for gas. Per unit of energy produced gas produces less carbon dioxide than oil does.

I was surprised he didn't slip in another excellent device for alleviating concern about global warming: the fact that carbon dioxide, the main greenhouse gas, traps progressively less heat as atmospheric concentrations rise. Sceptics tend to use this argument with abandon, but of course this aspect of the gas is factored routinely into the IPCC's calculations and estimates of what particular gas concentrations translate into what global average temperature and climate regime.

'I don't want to steal Jeremy Leggett's thunder,' the young chief geologist said. 'The difference between him and me is that he writes in the *Guardian* and I only read it.'

Self-effacing: a good tactic. It won him more laughter. Attending the lecture, I decided, had not been a good idea. My blood pressure was searching for the roof.

Having cleared the energy-policy opposition, the young oilman launched into a litany of success stories from Amerada Hess's casebook in the North Sea. Each was a high-tech hunt which had resulted in a kill. The students were shown the full gamut of million-dollar seismic profiles, intricate cross-sections of strata, well logs and geological maps, all by means of the gloriously designed slides which are the trademark of an oilman or oilwoman giving a public presentation. The oil industry hunts, and kills, with sophisticated toys. The students were scribbling. After all, it was very interesting. Finally, he finished with these words. 'Ladies and gentlemen, the world is your lobster.'

Apart from the disingenuousness of his treatment of energy policy, and his standard misrepresentation of the global warming threat, it had been a rather cruel piece of showing off. Nowhere had he mentioned that at that time only one in every ten geology graduates went on to careers in an oil company. The industry is high-tech intensive, but not labour intensive.

As the applause died down, I contemplated a cruel piece of showing off of my own. And I resolved to have no compunctions at all if it worked. As the lights went up, the President of the Geological Society called for questions. He saw my raised arm, recognized me in the same instant, and for a clear few seconds wondered if pointing at me was a good idea.

I tried to sound merely inquisitive. 'I was interested to hear that the life expectancy of the world's oil reserves has been extended in recent years. I was interested, too, in your comments about carbon dioxide and global warming possibly being a problem. I wonder, if we burn all the oil, and indeed the other fossil fuels, could you give us some idea how much carbon that would put into the atmosphere and how that would compare with the amount there at present? Then, given that carbon dioxide is a proven greenhouse gas, I'd be really interested to know how you figure there wouldn't be any global warming worth worrying about.'

He fell right into the trap. 'Er, of course, there is rather . . .' You could see on his face the mental scramble for an escape hatch. After a pause, he decided there wasn't one. 'Actually I don't have those figures to hand.' He waffled for a few more sentences, before deciding he was only getting himself into deeper trouble. Now all he could do was hope that I wouldn't follow the question up, and that the students had not attached any particular importance to it.

I fought to keep the anger out of my voice, but failed. All the oceans of my frustration, and my fear for the future, tended to empty themselves into moments like this.

'Well, I'm afraid that's just not good enough, is it?'

A forest of young heads turned to look at me.

'You don't know the basic arithmetic of the carbon cycle, yet you seek to persuade us that there is nothing to worry about in business-as-usual burning of fossil fuels. And that, I'm afraid to say, is symptomatic of your entire industry's response to the global warming threat. Let me tell you the statistics.'

I spelt it out. We knew from analysing ancient air in ice-cores that for thousands of years before the mass burning of coal, oil and gas began there were only around 580 billion tonnes of carbon in the atmosphere as carbon dioxide. Largely because of accelerating fossil-fuel burning since the Second World War, there were now more than 750 billion tonnes of carbon up there. We were currently adding 6 billion tonnes of carbon as carbon dioxide each year from the burning of oil, coal and gas. Ecologists were warning, meanwhile, that producing even 200 billion tonnes more carbon from the burning of fossil fuels risked ecological catastrophe. Beyond that point, the atmospheric concentration of carbon dioxide would be such that not only would the absolute temperature rise itself be a problem, but the prospect of natural amplifications of global warming would rise rapidly. Beyond the 300 billion tonne level, the risk would certainly become intolerable. Yet the total amount of carbon in fossil fuel deposits below ground was fully 10,000 billion tonnes, according to the best estimates at the time. Of that, some 200 billion tonnes was oil, almost 1,000 billion tonnes was gas and the rest was coal. Burning just a few per cent of this buried fossil fuel would amount to a flirtation with ecological catastrophe.

These figures for carbon in fossil-fuel deposits have been refined over the years, but the basic picture has not changed. As this book goes to press, the IPCC estimate for the total amount of fossil-fuel 'resource' – the quantity realistically recoverable, as opposed to total deposits – exceeds 4,000 billion tonnes, over three-quarters of it coal. To hike the global average temperature to a dangerous 2°C above pre-industrial levels, we would have to emit less than 220 billion more tonnes of carbon from fossil fuel, if the climate sensitivity is at the upper boundary of the estimated range, and little is done to stop tropical deforestation.

Of course, I told the Amerada Hess man, he could argue that we should cut down coal burning and favour oil and gas, and people did argue that. But even if we drastically cut back on coal use in the developed world, the amount of coal being burnt in China, India and elsewhere was rising quickly. There was no escape. We had no option but to leave the vast majority of the remaining fossil-fuel reserves in the ground: the vast majority of the coal, the great majority of the gas and much of the oil as well. All the billions being spent on continuing oil exploration should be spent on solar energy instead.

I had never tried to humiliate an opponent in this way before. And I never did get away with an intact conscience. As the session broke for lunch, the oilman came up to me, a sheepish grin on his face, and extended his hand.

'No hard feelings,' he said.

I didn't know whether he meant on his part or mine, so I made some inane remark about people just doing their jobs.

'You don't recognize me, do you?' he said.

And in that instant I did. He had been in the first set of students I had ever taught at Imperial College. He was not the chief geologist, of course. He had merely been standing in for his boss.

The idea for my cruel tactic with the carbon arithmetic derived from an event also connected with my former Mecca. Towards the end of my time as a geologist, I had been invited to join an institution known as the Geological Society Dining Club. The Dining Club was where the deans of the geological community met periodically, whether they inhabited the worlds of oil, mining or blue-skies research, to do what deans do over haute cuisine, fine wine, port and cigars: plot the future for their particular corner of the Establishment. I had been surprised indeed to be invited to join. This was a privilege usually extended only to geologists nearing the ends of their careers. I was barely thirty at the time. What was more, if even a tiny minority of the existing members were opposed when one of their number nominated a new member, they could stop the election of that person via a secret vote, a process known as blackballing. It was with some trepidation that I dined for the first time with the great and good of my profession. There I had been told, among the loosening tongues over the port, that I was the token young person in the club. The person who whispered this to me was, as she put it, one of the token women.

My membership of the Geological Society Dining Club did not terminate with my defection to Greenpeace, and one evening in April 1991 I went to a dinner to renew acquaintance with some of my old heroes. I sat with David Jenkins, a chief geologist at BP. I was in the middle of a convivial dinner, when Jenkins brought up my defection.

Why had I done it? He was intrigued. I had thrown away a promising career to join a scientific bandwagon and a bunch of rabble-rousers. So I rather reluctantly laid out my defence.

The BP man told me he could not understand why I was so worried. The danger, surely, was that oil and gas would run out before we could get alternative-energy technologies ready to take over, not that we would have to cut emissions from hydrocarbon burning ahead of that date.

I asked him how he could possibly hold such a view, given the arithmetic of carbon in the atmosphere and in the fossil fuels left below ground.

All of a sudden I realized that he had no idea what the sums were. I told him. David Jenkins listened to all this with a frown on his face. He had taken out a fountain pen, and was scribbling the numbers on a table napkin. 'Are you sure about these figures?' he asked.

I felt that I had discovered something very important here. The most basic information on the global warming debate, it seemed, was not getting through to people like Jenkins. I had had a similar experience at the World Climate Conference, where I had taken part in a public debate with BP's then managing director. It seemed that a kind of subtle corporate information shield was at work. People in the carbon-fuel industries were able to exchange perceived wisdoms about global warming in a comfortable, mutually supportive milieu into which few opportunities were offered for the insertion of worrying extraneous information. When information was aired widely in the industry, it was often spurious.

From the viewpoint of the environmentalist, the situation was at once depressing and yet intriguing. It meant that the power of reliable information could be strong indeed if one could just find the opportunities to get it across. I resolved there and then to do as much advocacy work within the oil industry as they would let me.

But why didn't the men from BP and Amerada Hess know the depressing arithmetic of carbon? The latter had taken a university science degree not long before. Indeed, I had been one of those who had taught him.

The answer lies in the culture. It isn't in the culture of a training for life in the oil industry – or within the industry itself, once you enter the fold – to deal with an integrated view of the atmosphere, the oceans, the biosphere and the entirety of the climate system. You would need a more holistic view of science, and the world into which that science has to fit, than the practitioners – academic or industrial – generally consider relevant to the needs of the trainee technician. The degree course the Amerada Hess man took, in particular, fell short on the atmospheric component. I taught on that course, and the shameful truth is that I didn't even know the carbon arithmetic figures myself – despite having done a Ph.D. at Oxford on ancient oceans – until the mid-1980s. Like my former student, I had no idea that there was so little carbon in the modern atmosphere, and so much in fossil-fuel reserves. Like him, I could not have guessed the quantities involved to within thousands of millions of tonnes.

DECEMBER 1991, SIBERIA

Through the ice-flecked window of a Tupulov airliner, shortly before Christmas 1991, I watched dawn break over Siberia. Peering down on a terrain which looked from this height as though it had been bombarded by meteorites and then frozen solid, I saw a long, narrow line marching dead-straight across the landscape. Then another, and another: tracks cut through the trees by seismic crews as they had flattened their way across the forest, blasting sound energy into the ground in order to map Siberia's oil-laden subsurface structures. Soon, the landscape was scratched by a network of these long-disused roadways. More or less evenly spaced, parallel to one another, with perpendicular cross-lines, it was as though someone had divided up a satellite picture of Siberia into an inaccurately drawn grid.

Then I saw the first oil well. Rather, it was a billowing flame, bright orange against the snow: gas being flared.

The stewardess's voice came over the intercom. 'We are about to begin our descent into Nizhnevartovsk,' she said in massively accented English. 'It is 33 degrees below zero.' The oilmen on board stirred, grumbling. The charter flight had been diverted here from Raduzhnyy. The men had already spent 24 hours sleeping in plastic chairs at an airport in Moscow while Aeroflot tried to locate some fuel for the plane.

I returned to my scrutiny of Siberia. We were over the heart of an oilfield. Rigs, almost all of them with burning gas flares, ran in clusters off into the distance, like torched Kuwait without the black smoke. Pale yellow smoke palls hung above the flames, motionless in the icy air. In the region below me, the Tyumen province of Russia, two-thirds of the former Soviet Union's yearly oil production was being pumped from the ground, at temperatures sometimes colder than minus 40°C.

Desperate for foreign currency, the disintegrating Soviet Union had effectively declared open season for Western oil companies in Siberia. And in consequence, Russia and the other former Soviet republics were on the verge of an old-fashioned, Western-style oil boom. I was flying in with instructions to take a look at the first outposts on the new frontier: to assess the damage that had been done by the Soviet oil industry, and to weigh the damage waiting to be done amid the boom anticipated by the new partnership.

The history of the oil industry is based around a sequence of oil booms. Such booms have been fuelled by the efforts of the industry's frontiersmen in ever more inhospitable conditions, in ever more remote terrain. The first was in Pennsylvania. Next, five years after Henry Ford built his first car in 1896, it was Texas. Just before the First World War it was Persia. Just before the Second World War it was the Arabian Gulf. In the 1960s it was the North Sea and Alaska. Now, on a planet running out of prospects for untapped oil provinces, and with Antarctica effectively declared out of bounds for many years by governments, Siberia was the new frontier.

There were at the time around 60 billion barrels of proven reserves of oil in the former Soviet Union, most of them in Siberia. For comparison, Saudi Arabia, since it entered the oil game in the 1930s, had by 1991 produced around 70 billion barrels. To the oil industry, then, the quantity of oil simply awaiting production is a treasure trove in its own right. However, on top of this, based on what is known of the unexplored terrain in Russia, the oil yet to be found runs to well over 100 billion barrels according to industry estimates. Together, proven and undiscovered oil reserves in the former Soviet Union approach 200 billion barrels. That is almost a third of all the oil ever burnt, globally, since the stuff was discovered more than a century ago. And then there is the gas. Here, the prospects are even more mouth-watering for the oil companies. The former Soviet Union has more than 40 per cent of the world's proven gas reserves, most of them in Siberia, and much prospective terrain is barely explored.

This is why no less than 36 companies had set up offices in Moscow by the end of 1991, at the time of my visit, and why, in eastern Siberia, the Japanese were casting covetous eyes over the potentially huge reserves in the largely untouched wilderness around Lake Baikal. The uncertainties of doing business in a country recently emerging from revolution, a country facing the prospect of civil war, and with an economy on the verge of total collapse – all that had been set aside in a renewed scramble for oil.

In Siberia during December 1991, one oil company was ahead of the game. Anglo Suisse had nothing to do with the English or the Swiss. It was a hard-nosed Texan outfit with a name deliberately chosen to convey an impression of sober neutrality. The day after my flight across Siberia's oil province, I stood with Ben Morris, an Anglo Suisse drilling manager, looking at a well just weeks from completion in the 350,000 barrel West Veryagan field. On the drill-rig platform, my companions on the flight in worked side by side with Russian roughnecks. A slogan on one of the hard hats, the motto of the drilling contractor, read 'Get Deeper, Quicker'. Behind us, in a sprawling nest of well-insulated prefabricated huts, off-duty drillers ate heaped plates of Texan food and watched videos. All of it, from the drill bit to the Coke machine, had been flown in from the USA.

'This is one impressive operation,' I remarked to Ben.

Ben nodded. 'While all the others're talkin', we're doin'.'

Along with a second field, Anglo Suisse had a controlling interest in half a billion barrels of oil in a joint venture with the Russians. The name they chose for it was the White Knights project. They were already lining up their second cooperation, an expected prize of over a billion barrels in just four fields. At that time, not even Russian bureaucracy was stopping them.

'We wanted to perforate a well,' Ben said, referring to the use of explosives down the well to improve the going. 'The locals said we couldn't. Too dangerous.' He shook his head at this misplaced excess of bureaucratic zeal. 'We showed 'em how safe the charges was, how many thousands of times they bin used elsewhere. They still said no. So we just done it. We told 'em, you better find a jail good an' big. And make damn sure it's well defended!' He laughed at the thought, and I laughed with him. Ben was a charismatic man, and great company. We were getting on well together, and I had to spend two more days with him in the oilfield. I imagined how disappointed he would be if he knew I worked for Greenpeace.

'What was the outcome?' I asked.

'They done nothin'. They'all about as pissed with their bureaucratic bullshit as us.'

So much for the chances for the embryonic oil-pollution legislation I had heard about in Moscow a few days previously.

The drilling rig, in the pitch black of early afternoon, was lit up like a funfair, billowing steam from an array of pipes as it drilled towards its target 3,000 metres below. It operated round the clock, seven days a week. Trucks and mobile cranes, parked on the drilling pad, pumped billowing clouds of exhaust from engines kept running continuously to stop them freezing up. The Russian and Soviet flags flew side by side on the drilling floor with the Stars and Stripes and the single star of the Texas flag. It was all as potent an image of the power of oil as I had ever seen.

Around the rig site were equally strong symbols of oil's wasteful use, and environmental toll. The dark sky above the forest was painted a flickering orange by the flares of the three producing wells in the vicinity. In the day I spent at the rig, the flares each burnt a hundred thousand cubic metres of gas. Like the oil itself would do, once burnt in far-off markets at the other end of the leaking Soviet pipeline network, the burning gas was releasing tonnes of carbon dioxide into the atmosphere each day.

I was travelling in Siberia with a film crew shooting a documentary on the oil industry for the American PBS and British BBC networks. Theirs was to be one episode in an eight-episode, Hollywood-financed serialization of Daniel Yergin's Pulitzer Prize-winning history of the hydrocarbon age, *The Prize*. It was to be the final episode, the one Yergin didn't actually write about – the future of the oil industry. Siberia, of course, was deemed by the producers of the series to be a big component of that future. I was in Siberia to be filmed talking to the chief of the local Khantsi people about the impact of the oil industry on their Inuit way of life. His was to be the local horror story, mine was to be the global. In a 50-minute film I would be lucky to be allotted more than a few minutes. Yet, for those few minutes I was having to spend four days in one of the most inhospitable terrains on the planet, a freezing genuflection to the importance of image in the television age.

Ben and his colleagues knew me only as a journalist, along for the ride to do a story on the filming, and hence writing down everything they said. The producer, Gregory Rood, had decided that an admission that Greenpeace was part of the film, even only in a walk-on part, would have risked his access. So I was to keep my mouth shut until he pointed a camera at me.

It was easy to see why. Ben's boss, the project director for Anglo Suisse, was a trim man called Gerry Walston. Gerry had a no-nonsense, military air about him. His employees were clearly in awe of him. Wherever we went, Gerry was with us. He was transparently keen that the BBC/PBS film should be as close to an Anglo Suisse public-relations production as he could get it.

After many delays getting through Moscow and into Siberia, we had only two days for filming, and Gerry had them all planned out for us. I watched an interesting battle of wills as Gregory manoeuvred delicately between his own film-making agenda and Gerry's. Gregory had one big problem: he needed access to a helicopter, and Gerry was the only man who could get it for him.

On the first day we shipped out to the West Veryagan field in a lorry. We left before dawn to maximize filming time. The city of Raduzhnyy, a blocky breccia of concrete in a matrix of snowdrifts and feeble neon light, was waking up as we rattled down the main road. Fur-swaddled figures stood in the dark on street corners, waiting for transport. All of the city – people, transport, concrete for the tower blocks – had been flown into Siberia, I was told. There was no road from Raduzhnyy to western Russia which could cross the huge Ob river.

We bumped along a 15-kilometre track through the taiga, a thin pine-forest terrain, to West Veryagin. There I learnt what I should have known already. The slow process of filming, a test of patience at the best of times, became a nightmare at minus 35°C. I was bundled up in the best kit my friends in the British Antarctic Survey had been able to lend me. But within 30 minutes I was casting envious eyes at the roughnecks heading for the cosy nest of prefabricated huts. After 40 minutes I left the film crew to it, and joined the drillers.

We moved on to the separation and metering plant. There, the oil was fed from black tanks into a pipeline. I had done a little research on Soviet oil pipelines. At the time there was more than 65,000 kilometres of crude oil pipelines in the former USSR, and more than 200,000 kilometres of gas pipelines. Most of these were in Siberia, where the difficulties of engineering, building on frozen ground which melts in summer, subsiding variably, meant that significant spills and leaks were commonplace. Specialists from the environmental protection department of Tyumen's Oil and Gas Extraction Authority, revelling in an opportunity to add to the catalogue of glasnost-shocks of the late Gorbachev era, had confessed that the loss of oil in the early 1980s amounted to as much as 1.5 per cent of the total extracted in the area. That meant up to 3 million tonnes per year, or more than 20 million barrels. Each year, in other words, 80 times the amount of oil spilt during the *Exxon Valdez* disaster was simply being shed into the soils, rivers and the lakes of the region. In the Tyumen area, when the snows melt in summer, vast standing lakes of oil can be seen. The biggest, at Samotlor, was 11 kilometres long and 1.5 to 2 metres deep.

Moscow news had reported in 1991 that, country-wide, 12 to 15 million tonnes of oil (88–110 million barrels, or between 350 and 440 *Exxon Valdez*'s) were spilt each year. A month before my trip to Siberia, I had heard Soviet Environment Ministry officials in The Hague report to the first international oil-industry conference on exploration, production and environment. One of the speakers had indulged in a spectacular session of glasnost-inspired soul-baring. He had spoken of the 'irreversible ecological damage' perpetrated by the Soviet oil industry in the Tyumen area.

One of the comforting thoughts on offer from Western oil companies at the time was that the Western oil industry could never make such an ecological mess as its Soviet counterpart had. But were companies like Anglo Suisse planning to engineer their own, safer, pipeline network? Anglo Suisse, Ben told me, paid the Russian pipeline company for use of the pipeline. They would be pumping their oil direct into the same old leaky pipeline network the Soviets had used all these years.

Gregory wanted to capture the flavour of this disastrous situation on film. His problem was that at this time of year there was nothing to see but snow. It was time to risk exposing my hand. I waited until Gerry was out of earshot and turned to Ben.

'Is there a place we can reach under the snow and find spilt oil?'

He was instantly wary. 'No, hardly any gets spilt.'

'You must have had some spillages, surely. We hear it can get pretty bad.'

'Only bin two pipeline leaks since I bin here. The Russians are real ecology-conscious. I'd say they have a better record than the domestic US industry.' Ben scrutinized me. 'Anyhow, the oil is a sweet light crude. It evaporates and breaks down.'

I nodded.

When oil is spilt in Siberia, the cold temperatures greatly slow bacterial decomposition. In many areas spilt oil simply sits there, seeping soluble, carcinogenic polynuclear aromatic hydrocarbons into the ground, lakes and streams, poisoning to varying degrees all creatures which live in, or use, the water – including humans.

Before dawn the next day, I was driven with the film crew to the airport, where a giant helicopter waited to fly us to a Khantsi village. The Khantsi's wooden huts were scattered among fir trees, a scene reminiscent of a film set for a Walt Disney epic. A crowd of about fifty Inuits, clad in colourful bundles of fur and fabric, awaited us, a forest of open Asiatic smiles on faces of all ages. The film crew went into barely suppressed rapture.

Lined up before the crowd of Khantsi were ten brand-new orange snowmobiles. Nearby, in a pile, were the packing cases from which they had evidently just been hauled. Gerry introduced us to the chief, Yuri, a balding man with an active manner and intelligent eyes. The filming began. Yuri gave a speech of welcome, translated for us by a Russian woman on Gerry's staff. Gerry gave a speech in return. The snowmobiles, he told the Inuits, were a symbol of the cooperation he hoped to have with them for many years to come.

The Khantsi smiled at him. One by one the snowmobiles were dished out. Yuri explained why each of the lucky individuals to be given one had qualified. You could tell why he was the chief. The first of them went to the poorest family, the second went to a family displaced from their traditional homeland where the oil now is. The politics was clear in each choice. Each of the lucky individuals gave a speech of acceptance, translated to a smiling Gerry.

Gerry was looking at the camera lens at regular intervals. I wondered what he would say if he knew that the cameraman had not turned his machine on. Gregory was running low on his stock of film.

A husky puppy, watching the bewildering proceedings, shifted one front paw off the ground, held it up for a minute, than alternated to the other, and repeated the cycle. The poor hound was as cold as I was. Soon the speeches became difficult to hear, amid crunching sounds, as human feet were lifted off the freezing ground. The recipients of the last few snowmobiles had to compete with the noise of engines being started, and the excited cries of children as they piled on for their first ride in the new family sledge. The smell of partially combusted hydrocarbons was soon thick in the air.

Gregory now set about shooting what he told me was to be my entrance in his film. I had to drive a snowmobile up to the door of Yuri's hut, as though visiting him. I had a three-minute lesson in how to drive a snowmobile. I set off as directed, half a dozen laughing Khantsi children clinging to the back of the machine. The cameraman sat on the back of another snowmobile, ahead of me. I made to follow his snowmobile around a corner. My machine sailed straight ahead. It rammed into a fir tree. Driver and Khantsi kids went flying into the snow.

I dug my way out, to howls of laughter from the watching Khantsi, and a somewhat strained reaction from Gregory, increasingly worried about his film supply. We finally entered Yuri's hut. Inside, a log fire spat fragrant sap. I sat among the reindeer rugs as Yuri's several wives served tea. Space was limited, and here Gregory seized his opportunity. Everyone would have to leave, he announced, for the filming of his interviews with Yuri and Jeremy. Including Gerry.

It worked.

The firelight flickered on the log walls. Yuri was known to be a poet. Gregory encouraged him to read a poem he had written about the poisoning of his land by the oilmen. Yuri looked uncomfortable, and volunteered instead to read a poem about his daughter.

'Why not read them both?' Gregory suggested.

Yuri agreed.

He read, in a mournful voice. He was a natural actor.

'This is a film-maker's dream,' Gregory breathed to me.

The film crew had interviewed Gerry at the West Veryagin rig the previous evening. I was to see what he said, on British television, a year and a half later. By that time, *The Prize* had already become the third most watched TV documentary series in US television history.

'We're the only ones here because we were the first to the plate to do it. Everyone else was cautiously watching. Many other oil companies are going to follow behind us.' Standing in front of his steaming rig, Gerry had a triumphal air. 'It's the unleashing of perhaps the next capitalistic frontier in the world, and certainly the unleashing of one of the great remaining oil provinces to be fully developed.'

Gregory told me at the time that he had also asked Gerry about climate change. Not a problem, Gerry had said. Climate is something you measure over decades. How could anyone know it was changing?

Were there not tremendous environmental pressures on the oil industry now?

True, Gerry conceded, the major oil companies were under siege from the environmental lobby at present. But the majors have lost touch with what their main mission is. And that was to find oil and make money.

A year or so later Gregory Rood called me when he was in London. He told me that his first cut of the film had had a heavily environmental slant. The series producers had disliked it. Gregory had been sent back to the cutting room. The series producers had not liked the fruits of his revision either. The film had been taken out of his hands.

Gerry's thoughts on climate were missing from the film as finally edited, along with much of the other environmental footage from the original director's cut. The theme was very much that of Siberia as the new frontier. Gerry Walston described Siberia in those terms on camera. Daniel Yergin, author of the book on which the series was based, backed this up. To this day, Gregory Rood has not seen the version transmitted.

A short portion of the interview Gregory Rood had shot with me in Yuri's hut survived. I was to be seen talking about greenhouse gases and the 'dangerous game of roulette' we would be playing if we burnt even a fraction of the world's oil reserves. I talked of jeopardy to the 'environmental security of generations to come' and the 'race against time' to force the oil industry to recognize a new energy future.

I looked cold. I sounded incongruous.

Sir Peter Holmes, chairman of Shell, was filmed in his warm and spacious office. He argued otherwise. For him, there would be no solar century, just another oil century. 'There really isn't an alternative, so far anyway, to the internal combustion engine,' he reasoned. 'Oil and gas will be major industries fifty and probably a hundred years from now.'

DECEMBER 1991, GENEVA AND THE HAGUE

I flew out of Siberia, via Moscow, to a Geneva decked with Christmas lights. But in the Palais des Nations, at the fourth session of climate negotiations, no festive spirit was evident. Discussing commitments one day soon after I arrived, the negotiators spent 165 minutes over one paragraph of 109 words. That worked out at roughly one and a half minutes per word. By the end, the chair had suggested putting together a sub-group to synthesize the different proposals on the paragraph. But, observed the Algerian delegate, there were in all 71 paragraphs for negotiators to consider. Did that mean there would have to be 71 sub-groups?

At an evening reception, I found myself talking to a Kuwaiti and an Iranian, the latter representing the OPEC secretariat. Were they not worried that this year had been the second hottest on record, I asked (this had just been confirmed by the UK Met Office, notwithstanding the six months that had been cooled by Pinatubo's eruption), and that the year before was the hottest ever?

They came from countries that were used to heat, they told me.

We passed through the usual conversational foothills, and then got to the customary mountain. 'But are you happy,' I asked, 'to take all that carbon up from the ground, and add so much carbon dioxide to an atmosphere which, for so long, has been accustomed to so little? Can you be sure that if we do that, we won't wreck the entire planet?'

'Can you see it?' the Iranian asked.

'I beg your pardon?'

'Can you see this carbon dioxide?' He gave me a little smile that suggested he thought he had made a telling point.

I was out of my depth here.

Dan Reifsnyder of the State Department drifted into the group. Since our last conversation, John Sununu, the Bush White House chief of staff, an unreconstructed opponent of greenhouse-gas limitations of any sort, had been forced to resign.

'Will it make a difference to the policy?' I asked Reifsnyder.

'Well, some think so. But I think,' he paused for effect, 'never.'

While we were in Geneva, Western Samoa was hit by the worst cyclone in Pacific history. It turned my colleague Pene Lefale's homeland into a disaster zone. At least seven people were dead. Pene was waiting for a plane in Auckland to get home. He did not yet know whether any members of his family were among the dead or injured. There had been fatalities in both Western Samoa and American Samoa, which President Bush had declared a disaster area. Thousands had been made homeless as entire villages were destroyed by four days of winds up to 300 kilometres per hour and waves up to 18 metres high. It had become clear from the early reports of New Zealand relief teams that Cyclone Val was worse than Cyclone Ofa, which hit the Samoas early in 1990. Ofa held the previous record, for both intensity and damage: a bill of $160–170 million.

Pene had to wait for several days to get into Samoa. During that time, he saw TV pictures that panned round his flattened home town to within one house of his own. He still did not know whether his family were among the dead and injured. Only when he finally arrived did he find that they had all survived. I knew Pene like a brother by this time. As though his commitment to fighting for the Pacific island nations had not been strong enough already, this disaster carried it into a wholly new dimension. From this point on, whenever he and I worked together lobbying foot-draggers of all complexions – as we frequently did – I would wonder how on earth he managed to maintain his statesmanlike calm.

At this stage, the churches made their move at the negotiations. The World Council of Churches, representing millions of Christians in over a hundred countries, issued a statement in Geneva urging the industrialized nations to act. 'Concern for the protection of Creation is growing rapidly in the churches,' it read, 'with global warming being a high priority.' There was an urgent need for the industrialized nations to set specific targets and schedules to reduce emissions of greenhouse gases.

But the next day, Bob Reinstein reaffirmed the US opposition to targets. The head of the US delegation, in a radio interview earlier in the week, had described the EC's commitment to targets as 'purely political'. But the EC was attempting that very week to move from political commitment to policy substance, and in so doing was providing the only crumbs of comfort to be seen on the climate horizon. A Council meeting of both environment and energy ministers in Brussels instructed the European Commission, the EC's executive arm, to draw up a draft directive for an energy tax, for prospective agreement before the Earth Summit. The proposal was for a tax which would start at $3 per barrel of oil (or oil-equivalent), rising to $10 per barrel by 2000. Reuters reported that Germany, the Netherlands and Denmark were the main supporters. The UK, as usual, was dragging its feet.

Senator Al Gore, the champion of US action on climate, now flew to Geneva to do what he could to draw attention to the flagging negotiations. He gave his own rendering of President Bush's track record to a packed meeting of NGOs. 'When the history of this period is written,' he said, 'I believe this may be seen as the single worst abdication of leadership ever.'

Paul Hohnen and I spent an hour in the Palais with Gore and his staffer, Katie McGinty, later to be President Clinton's environment advisor. Much of our conversation concerned the Arctic. Gore had travelled there with the US Navy. We talked through the prospect that the Arctic ice cap was already thinning, and the implications of that dreadful possibility. Gore was one of the few politicians I had ever met who was aware of the scope for large-scale flips in ocean currents, or had heard of methane hydrates, and knew that warming Arctic waters, and tundra, could start releasing huge quantities of methane.

He had written a book, he told us. It came from the heart. He had written every word of it himself. It was called *Earth in the Balance*.

The environment groups worked hard on a pooled statement for the final plenary session. It was given by Dr Michael Oppenheimer, the Environmental Defense Fund's star climate scientist. The American spoke for thirty environment groups on six continents, and he did not mince his words. The delegates fell silent as he spoke.

Oppenheimer's voice shook with emotion as he came to the role the US delegation had played. 'My country – that renegade country, the largest greenhouse polluter in the world, the world's wealthiest country – has yet to commit itself to doing anything.' He went on to the threat. He had been a world-class astrophysicist before joining the environmentalists, and was a reviewer for the IPCC scientists' working group. Everyone in that hall was aware that he knew his stuff. 'We are sure that large and potentially disastrous changes are afoot. Let us be under no illusion. This is a consensus among scientists.'

Then he came to Saudi Arabia and Kuwait. 'Their commitment to oil,' he declared, 'outweighs their commitment to the fate of humanity.'

The fourth session of negotiations broke up just before Christmas 1991 in a state of disarray.

3

The Road to Rio

January–June 1992

JANUARY 1992, GUANGZHOU, CHINA

Exhausted, I put off 1992 until 11 January, when I had to board a flight from London to Hong Kong. It was an airline I remembered for its good food. This time around, however, a slug crawled from my salad across the table-tray towards the draft of the Intergovernmental Panel on Climate Change's 1992 science update report, which I was in the process of reading through a permanently twitching right eyelid – stress symptom of the month, it seemed. I handed this rather bad omen on a napkin to a Chinese air hostess. She examined it as though assessing which species it belonged to, and I saw that my plan to angle for an upgrade to first class would fail.

From Hong Kong, I flew across the Pearl river delta to Guangzhou, and the five-star Dong Fang hotel with its international convention centre. I had no idea that such places existed in China. My recollections of the country from the early eighties were more of breeze blocks and brick dust than of plate glass and thick-pile carpets. The Dong Fang's two huge buildings faced each other across an oriental garden in which palm trees lined contoured lawns, and lakes – more goldfish than water – wound under ornate footbridges.

Having attempted in vain to strike up a preferential relationship with the operator of the single fax machine, which I knew would be red-hot before the IPCC meeting was over, I dined alone with the latest John Le Carré. At the next table, ICI's representative was attempting to persuade three climate scientists that there was no alternative to HFCs for refrigeration. HFCs were the chemical industry's gas of choice to replace ozone-depleting CFCs as the coolant for fridges. Although they do not deplete ozone, HFCs are potent greenhouse gases. I ate my eels au Fang, gratified to overhear that the ICI man's arguments were being received with a degree of scepticism.

The next day, I arrived at the conference hall early. I took my place at the back, at the table for NGO observers. The environment groups had only managed to send two observers, and I was the only scientist. Fatigue was affecting not just the campaigners, but the budgets, it seemed. The carbon club was represented in force, with both scientists and non-scientists. Don Pearlman and his cronies were shooting for a big prize here in China. If they managed to water this report down, such impetus as existed would be completely sucked out of the climate talks.

John Houghton, chairman of the working group, called the meeting to order. He was pleased to welcome the hundred-plus scientists to China, he told them. They represented, along with the reviewers not present, over forty governments, a reflection of the international interest in the work of the IPCC Scientific Working Group. China was a very appropriate location for their deliberations. One reason was the dependence of China's economy on the climate.

In particular, I pondered, its agro-economy. The previous year had offered Chinese farmers a suspiciously atypical and challenging set of extremes. In May, a long-running drought, delaying the spring planting, had threatened the Chinese rice crop. Then in the summer came the floods of the century in the Yangtze valley. And there followed a three-month drought covering an eighth of the country, and culminating in November in a devastated grain crop.

'Well over 150 scientists have contributed to this latest draft,' Houghton continued, 'some even working Christmas Day. The integrity of the science is absolutely vital to the success of the IPCC. We must not allow ourselves to be influenced by politics.'

I glanced at the hard faces along the row of my fellow observers. Two new developments in the science had given them plenty of good material with which to make mischief if they wished. First, a panel set up under the Montreal Protocol to review ozone trends had made the remarkable and alarming discovery that CFCs were now depleting ozone so unexpectedly quickly that the heat-trapping properties of CFCs were effectively being cancelled out by ozone loss. This meant that CFCs were less important a contributor to the human-enhanced greenhouse effect than had been thought by the IPCC scientists at the time of their 1990 Scientific Assessment Report. The overall rate of global warming would be suppressed, in consequence. Second, new work on sulphate aerosols – produced naturally when volcanoes erupt and unnaturally when coal and oil are burnt – suggested that they were more important than had been thought in 1990. Before they were incorporated in acid rain, the tiny aerosol particles spent time reflecting a lot of the sun's rays back into space. This meant that there were now two ways in which the overall human-enhanced greenhouse effect was being damped.

There was nothing whatsoever to be pleased about here. CFCs had to be abandoned as soon as possible anyway, unless we wanted to continue stripping back the life-protecting ozone layer. Sulphate aerosols, if not as deadly as CFCs, were highly dangerous once rained out of the atmosphere as weak acid, unless we wanted to continue killing forests – forests which among other things were capable of taking carbon dioxide down from the atmosphere. Simply stated, we could not go on merely suppressing a fever with two drugs of such toxicity that they themselves risked crippling the patient.

A hundred documents rustled as the meeting's attention turned to the draft report.

'Findings of scientific research since 1990,' the critical statement in the summary read, 'do not affect our fundamental understanding of the science of the greenhouse effect and either confirm or do not justify alteration of the major conclusions of the first IPCC Scientific Assessment.' In other words, the threat is just as real as it was two years ago when the IPCC's first scientific assessment kick-started the negotiations for a climate convention, even though detailed understanding has been evolving. If the industry representatives succeeded in forcing a retreat from this conclusion, they would score a major victory.

The carbon club reserved its attack on the bottom-line statement for late in the meeting. It was not Don Pearlman, but a newcomer from the Australian Coal Association, David Hughes, who fronted the bid for a home run.

'Given all the uncertainties over estimates, based on ozone depletion and sulphate aerosols suppressing warming and the rest, surely we can no longer justify this statement, Mr Chairman.'

'This form of words has been commented on by many referees,' John Houghton said stiffly.

Just like Exxon's Brian Flannery at the key IPCC scientists' meeting in 1990, Hughes found no support outside the carbon club.

I flew back to Hong Kong and spent a day in a hotel in Kowloon, faxing and phoning, pushing the report out to waiting environmentalists and key journalists. The meeting had been far from the disaster it could have been. The 'authoritative early warning' of 1990 had been reaffirmed. But the message of the 1992 science update was now more nuanced than the conclusions of the first report had been. Because the dampening effects of ozone depletion and sulphate aerosols varied enormously around the globe, the IPCC had found difficulty estimating by how much the two processes would reduce model-based estimates of global warming, other than to say it would be by a small amount. The UK delegation settled on a crude estimate of 20 per cent. This reduction would send a message to the world, which would inevitably be interpreted almost everywhere as less cause for concern.

Nothing was further from the truth. The revised rates of warming left model-based estimates closer to the observational record, so increasing confidence in the forecasting ability of the models. The lower predicted rate of warming, still well over 0.2°C per decade, remained above the threshold of ecological danger, which many ecologists placed at 0.1°C. The potential for a worst-case analysis of synergistic positive feedbacks still remained buried from view. There had been no specialist ecologists at the Guangzhou meeting. All the key players had been atmospheric scientists of various kinds.

I flew back to the UK, and scoured the newspapers. 'ESTIMATES OF GLOBAL WARMING SCALED DOWN,' read the headline in the *Financial Times*.

FEBRUARY 1992, TALAHASSEE / WASHINGTON, DC / NEW HAMPSHIRE

We were now into the Earth Summit endgame, and its location was appropriate. From the end of January until I left for Brazil on 11 May, I was to spend all but a few days in the United States. The final session of the climate negotiations prior to the Earth Summit took place at the UN in New York in mid-February. The USA, having hosted the first session in the backwoods of Virginia, had fought to keep subsequent sessions of the negotiations abroad. In New York, whatever happened, there would be some attention in the US media. And where there was media attention, especially during President Bush's campaign for re-election, there was the possibility of embarrassment for the government, and hence a shift in policy.

I flew first to Florida. It had not escaped the attention of some Floridians that in most places around the Florida coast, you had to go a long way inland before you could find land a metre above sea level. In coastal communities, homeowners were beginning to wake up to the threat from rising sea levels and amplified hurricane activity. Tourist boards were becoming increasingly worried about the decline of the reefs, and the reports of coral bleaching. Water authorities were beginning to worry about the threat of aquifers becoming salinized. In December 1991, an opinion poll in Florida had shown that 71 per cent of those canvassed believed global warming to be a serious problem. A group called Citizens for the Preservation of Florida was staging an ambitious conference on climate change, hoping to push the issue onto the agenda in the primaries.

The contenders were beamed by satellite into the conference to explain their policies on global warming. Several hundred Floridians watched as Bill Clinton appeared on screen. The USA should at least freeze carbon dioxide emissions, like the EC, the future President professed. And a 20–30 per cent cut may even prove possible. The Floridians applauded.

Al Gore dropped into the conference to give a speech. He had just come back from the negotiations in Geneva, he told the audience. As a red-blooded American, it had all been a deep embarrassment to him. He came from a state with a hundred thousand tobacco farms, he said. His sister died of lung cancer – he held her hand while she breathed her last. When he heard scientists from tobacco companies questioning the link between smoking and cancer, he got angry. They should hang their heads in shame. It was the same thing with this issue. 'We have the equivalent in our civilization of a ten-pack-a-day habit,' Al Gore said indignantly. The Floridians gave him a standing ovation.

I flew from Florida to Washington, DC, and went straight from the airport to a press conference at NASA's headquarters. The latest findings on ozone depletion in the Arctic were to be announced, and the word was that the news was bad. By the time I arrived, 15 TV crews, including all the networks, were obscuring the view of most of the journalists crammed into the press room. Three scientists and a NASA official sat at the head table, waiting grim-faced for the signal to start.

Then the bad news began. The chief scientist described how NASA's converted spy plane, flying north into the Arctic, had – while it had yet to clear US airspace – encountered levels of the ozone-depleting chemicals higher than any measured before, at any place in the world, including the Antarctic ozone hole. Sheets of chlorine monoxide, the CFC-derived precursor to ozone depletion, were extending south out of the Arctic over populated areas of North America.

The cameras rolled as the scientists answered a barrage of questions about skin cancer, potential weakening of the human immune system, threats to plants, threats to phytoplankton, and the rest. NASA's men were sober, hesitant, tangibly shaken by the story they had to tell.

The next morning, both the *Washington Post* and the *New York Times* ran appropriately blistering lead editorials. 'Once again, it turns out that the protective ozone layer in the sky is being destroyed faster than even the pessimists had expected,' observed the *Post*. 'There's an important lesson in that bad news, and it applies to the much larger and more difficult question of global warming.'

I flew next to New Hampshire, to join my Greenpeace USA colleagues in an effort to inject global warming into the primary. A week of spectacular failure ensued. An ozone hole may have been in the process of appearing over President Bush's home in nearby Maine, but he was not in the mood to talk about it. With the exception of Governor Gerry Brown, neither were any of the other candidates. Nor could the campaign-trail journalists be persuaded to follow up on the editorialists at the *Washington Post* and *New York Times*, and raise the environment as an issue.

Back in New York for the next set of climate negotiations, I dropped into a bookshop on Fifth Avenue, looking for a copy of Al Gore's book, *Earth in the Balance*. I asked the shop assistant where the environment section was located.

'No special section, sir. It's under consumer affairs, nature, and trivia.'

FEBRUARY 1992, NEW YORK

There were now only a hundred hours of negotiating time left if governments were to complete the climate convention by the time of the Earth Summit. A hundred hours in which to negotiate the bulk of one of the most complex multilateral treaties in history.

As they settled down to the task in the UN building on First Avenue, rumours were rife in the coffee bars and corridors. Sources in the European Community now said that they thought the US position was about to crumble. But the hardliners on the US delegation asserted precisely the reverse, that it would be the Europeans who would buckle at this session.

At the first evening reception of the session I sensed that President Bush's foot soldiers were beginning to feel a degree of pressure. The reception was being hosted by AOSIS. The island alliance had gone to the effort of providing typical Caribbean and Pacific island food for the delegates, and music to match. They were still hoping the men from the White House might prove to have hearts that were reachable.

'Island culture,' I said to Dan Reifsnyder, motioning to the laden tables. 'Enjoy it while you can.' I smiled to soften the barb.

The State Department's number two at the negotiations proved to be in reflective mood. 'You know, Jeremy, I sometimes sit with one of my colleagues and ask, "what if we are the bad guys?"'

I put down my plate. 'Dan,' I said. 'Listen. I'm sure you love your kids just as much as me, but I'm telling you, I'm *certain* you're the bad guys.'

Reifsnyder looked at me with a blank face.

Events in the Senate that day precluded an overambitious flirtation with hope. A bill on energy, the Johnston–Wallop Bill, had been passed. It was a ragbag of lost opportunities and measures pandering to America's continued addiction to fossil fuels. Among these, electricity generation had been deregulated in a manner which would foster increased coal use, and worst of all, a proposed amendment to raise automobile efficiency standards to 40 miles per gallon had been withdrawn. That single provision could have saved two and a half million barrels of oil a day – fully half of all US oil imports from the entire Arabian Gulf.

But the bill was not all bad news. There were some measures to increase the efficiency of lighting, motors and commercial air conditioning, and a contentious proposal to drill for oil in the Arctic National Wildlife Refuge had been dropped. But the automobile and fossil-fuel industries' lobbyists in Washington had clearly been doing their job well.

The Global Climate Coalition was at the climate negotiations in force. As of January, they had a new executive director, John Schlaes, who was to become quite as visible as Don Pearlman in the corridors of the UN. Schlaes had been Director of Government Relations for the Edison Electric Institute before taking up the Global Climate Coalition job. He was a man who knew about influencing governments, and he brought with him a particularly bullish style.

The Coalition opened the batting with a pre-emptive strike the day before Bert Bolin was to report to governments on the 1992 IPCC Science Assessment. Schlaes organized a press conference at which Professor Fred Singer and another well-travelled sceptic attacked the IPCC's version of greenhouse science. Singer, a University of Virginia scientist and ex-science advisor at the US Department of Transportation, told the press that there was essentially no scientific evidence for global warming.

Singer had become well known for his performances at a series of set-piece conferences arranged by an anti-intervention industry front-group called Consumer Alert. At these events, Singer had variously professed not only that global warming was a myth, but that acid rain was an illusion, and that there was no proof that CFCs depleted ozone.

The Global Climate Coalition's lobbyists now released single-page briefing sheets for delegates at the climate negotiations in which they nailed their true colours to the mast. The stakes at the negotiations had become high, and Schlaes had evidently decided that there was no longer any need to pull punches. 'Stabilizing carbon dioxide emissions,' one of the briefings was headed, 'would have little environmental benefit.' The scientist cited in the few sentences offered in support of this exhortation to burn away was Professor Richard Lindzen.

The World Coal Institute, not to be outdone, flooded the negotiations with broadsheets, modelled on the environmental NGOs' newspaper, *Eco*, and entitled *ECOal*. 'The benefits of increased carbon dioxide have been ignored and the warming exaggerated,' read one article.

It was as though, with the faint prospect of a meaningful climate convention now on the table, the oil and coal lobbies had dropped all pretence at finessing their message. They were appealing to the baser side of the delegations from nations with particular interests in fossil fuels, whether as producers or consumers.

In the early days of the supposedly final session of negotiations, there was no sign of a US change of heart. Summing up the progress, the chairman observed that there were almost as many brackets left as minutes to negotiate them. The US meanwhile argued for 'voluntary commitments' for stabilizing greenhouse gases.

At this stage, I made available to the negotiators and press at the UN an opinion survey of the IPCC's scientists, in which I had tried to tease out the worst-case analysis. I had polled 400 climate scientists during December and January. The sample included all scientists involved in the 1990 IPCC study, and others who had published on issues relevant to climate change in the international journals *Science* and *Nature* during 1990. By the end of January, 113 had replied; 15 had professed that the runaway scenario was probable, 36 that it was possible, and 53 probably not. In other words, 51 of the 113 believed the runaway greenhouse effect to be at least a possibility.

Nobody who read the IPCC's science reports of 1990 and 1992 could ever have guessed that there were scientists out there in the world who believed that the worst-case analysis could reveal a point of no return. Those who believed a runaway greenhouse effect probable – barring cuts in emissions – included the director of the Max Planck Institute for Meteorology in Hamburg, Germany; the head of the German Meteorological Observatory in Potsdam; and the head of Institute of Astronomy and Geophysics in Louvain, Belgium. Those who believed it possible included recognized authorities from the Royal Dutch Meteorological Institute, Australia's Commonwealth Science and Industrial Research Organisation (CSIRO) and the Max Planck Institute.

I had released the survey earlier in the month, at the American Association for the Advancement of Science's annual meeting in Chicago. I had flown there from Washington immediately before the New Hampshire trip. The *Boston Globe* had picked it up and run a major story. I now left 400 copies of the *Boston Globe* article at the back of the hall in the UN. They disappeared almost at once.

Word soon came from within the Japanese delegation that representatives of the Ministry of International Trade and Industry had little perception that the ultimate stakes with global warming could, in principle, be so high. Finally, I felt, I had found an effective way of bringing the worst-case analysis to the attention of the negotiators.

The day after our press release, I had a call from CNN. Would I take part in a TV debate? It was with Professor Fred Singer, the producer said, an interesting scientist who didn't think there was a problem with either global warming or ozone depletion.

I took a deep breath. Did he realize, I said, that this was a man who had a long career of spouting the most spurious arguments of the most regressive quarters of the business lobby? A man who was flown around the world by the coal industry to offer an opinion contradicted by one of the most exhaustive international scientific consultation processes ever? I found it bad enough, I said, that with global warming, TV constantly picked scientists from the 1 per cent fringe and pitted them against a representative of the 99 per cent majority. But with ozone depletion, where we can be as certain of proof as we could ever be, this was like saying the earth was flat.

'Hey,' said the producer. 'That's great. Can you say all that on air?' It would make good television, he explained.

Two days later, I found myself sipping a coffee next to Saadalla Al Fathi, Head of OPEC's Energy Studies Department, and their observer at the negotiations. He had replaced the Iranian who had no worries about carbon dioxide because he couldn't see it. Al Fathi and I fell into conversation. He was a pleasant man. We found common ground: Imperial College had rejected us both as undergraduates. He had gone to Manchester, instead, to do engineering. I told him that Imperial had subsequently made the mistake of hiring me on their faculty. He hadn't missed much, I assured him. Then we found ourselves discussing the business at hand, where it was clear we were not going to find the identification of common ground quite so easy. As he must know, I said, I was of the view that to continue burning oil at the volumes we do today was far too much of a risk to a viable future. Why couldn't OPEC accept a much higher price for oil, for less production, I asked.

But how, he asked. It just wouldn't happen. And as for the risk, well, lots of scientists said there was zero risk.

With great respect, I replied, that was not my appreciation of the predominant scientific view. Wasn't he worried by what Professor Bolin and the IPCC had reported?

He wasn't. He felt the need to lay out his entire case, and told me what I knew all along, but had never heard expressed directly. 'You ask me why I'm here? I'll tell you. My motives are selfish. We don't want this convention. There's nothing in it for us.'

I excused myself, with a heavy heart.

Worst-case survey notwithstanding, the fifth session of talks went dreadfully. I sat disconsolately at the back of the negotiating chamber, watching the early hopes of a change of the US position evaporating. In the corridors and coffee bars, day by day, I saw frustrations rise and tempers begin to fray, even among seasoned diplomats. It became clear that substantive progress was impossible, and that yet another session of negotiations – on the eve of the Earth Summit itself – would be needed if a climate convention was to be ready for world leaders to sign in Rio.

Late in the session, the USA launched a diversionary tactic. Bob Reinstein announced that the White House would be unilaterally contributing $50 million for developing countries to conduct studies of their own greenhouse-gas emissions. For a New York press corps hard pressed to follow the significance of detail in these negotiations, it was just the job. It offered the sense, for the uninitiated, of a progressive US position. It diverted attention to the growing emissions of developing countries, and away from the USA – the source of a quarter of all greenhouse gases emitted.

I left New York knowing I would have to come back to those soulless UN chambers just a few short weeks later, once again to watch the cynical diplomatic games, still not knowing whether, at the Earth Summit, the world would have a Convention on Climate Change, or even, beyond Rio, an agreement to keep talking about the global warming threat.

APRIL–MAY 1992, NEW YORK

On the eve of the last-ditch session of climate talks, an Oxford University report for the United Nations pointed out that if the latest models from the UK Met Office and NASA's Goddard Space Flight Center were accurate, most of the world's important food crops would be hit by global warming. If global warming progressed in the way the models suggested, there would be a decline of between 10 and 15 per cent in grain yields in Africa, tropical Latin America, and much of India and South-east Asia. One in eight people could face famine within fifty years. Another report released during April, by the World Health Organization, concluded that global warming could cause the spread of malaria and other tropical diseases to millions of people presently free of them. Meanwhile, in drought-stricken Zimbabwe, game wardens announced that 90 per cent of the wildlife in the south of the country faced death. In a previous bad drought year, 1983, 80 per cent of the game animals in the Gonarezhou National Park had died. Then, there had still been some surface water left, this time, there was essentially none. Neither were the problems limited to animals. In Zimbabwean cities, riots were breaking out in food queues.

In the USA, the White House finally came under pressure. In the House of Representatives, the chair of the Subcommittee on Health and Environment, Henry Waxman, drew up an act which sought to commit the US to freezing carbon dioxide emissions. He went to the House for bipartisan support, and found it. Waxman, a Democrat from Los Angeles, called his bill the Global Climate Protection Act. By 7 April, the list of co-sponsors had grown spectacularly. The White House's Council on Environmental Quality decided to write to every congressman and congresswoman to try to stop the rot. Waxman's assertion that US negotiators had jeopardized the chance of achieving international agreement on climate change, they said, was patently false. President Bush's men urged Congress not to support the bill.

Waxman responded spiritedly on 15 April. Where, he asked, was their evidence? At a recent briefing in Washington, a senior EC official had characterized US opposition to stabilization of carbon dioxide emissions as the central obstacle at the climate talks. Diplomats from Canada, Japan and the Netherlands had agreed. 'I am pleased to note,' Waxman wrote, 'that despite the claims in your letter, and despite ludicrous claims of massive economic impacts from industry opponents, a bipartisan coalition of nearly 150 congressmen have endorsed the Global Climate Protection Act.' The same day, Waxman wrote to the president telling him of the support he had found in Congress, and urging him to re-evaluate his position.

On 22 April, sensing that attention on the issue was growing, Bill Clinton – now emerging as the clear Democratic front-runner – used Earth Day to step on the gas. 'Our addiction to fossil fuels is wrapping the earth in a deadly shroud of greenhouse gases,' he said in a speech in Philadelphia. 'Our air will be more dangerous because George Bush put Dan Quayle in charge of the Competitiveness Council, a group which lets major polluters in through the back door at the White House to kill environmental regulations they don't like. And the most disturbing thing is, they call that competitiveness.'

Bush Administration officials now submitted an interesting document, entitled 'US Views on Climate Change', to a meeting of OECD countries in Paris, where governments were seeking a way forward ahead of the last-gasp session of climate talks. A section on science, representing a consensus view of a broad range of scientists, 'including most US scientists', concluded that 'the best scientific information indicates that if greenhouse-gas concentrations in the atmosphere continue to increase as a result of human activities, significant changes in the climate system are likely.'

The sixth and final session of negotiations opened at the UN on the morning of 30 April. With Paul Hohnen, I took the familiar walk down 42nd Street. This time, we waited on the Manhattan side of First Avenue, facing the edifice of the United Nations, and with our backs to the towering UN Plaza skyscraper.

A yellow truck and a limousine swung out of the traffic, and parked across the entrance of the United Nations. Boiler-suited people leaped from the back of the truck. The limousine was driven by a man wearing a George Bush mask. The driver pulled a flap back to reveal a painted slogan across the side of the truck: 'Bush Blocks Climate Treaty. Danger. Global Warming Ahead.' People scurried beneath the truck and the limo, and began chaining themselves to the wheels of the two vehicles. Others headed for the fence in front of the UN's many flagpoles, carrying ladders. TV camera crews appeared as though from nowhere, some dodging through the traffic on First Avenue, others already filming the activities by the blocked gate. Blue-uniformed UN security guards came running from the UN building towards the scene. The ladder-bearers threw their ladders against the fence, in synchrony, and held them steady while others, draped in climbing gear, ran up the rungs as though they were commandos training and began to climb the poles.

The traffic slowed. Motorists craned to look.

'They won't even put them in jail!' a disgusted Buick driver called to the gathering passers-by.

'They should put George Bush in jail,' a passer-by yelled back at him. Office workers craned from windows across the avenue, clappers competing with hooters.

The climbers were now at the top of the poles, and seemed to be in the process of chaining themselves there. The ladder-bearers had immediately withdrawn, and several dozen security men were now gathered around the poles, shouting at the climbers and pulling roughly on the ropes. A woman climber was pulled down into the mêlée of gesticulating blue uniforms.

The TV cameras rolled. Sirens now filled the air. Half a dozen police cars arrived at the same time, parking at all angles. Their occupants rushed out into the crawling traffic and shut off First Avenue. A police helicopter swooped across the Hudson River, and hovered over the street. Now a fresh disturbance arose. A crowd had gathered on the far side of the street and were looking high up the towering side of the UN Plaza. There, two climbers had begun a slow descent on ropes from the dizzying height of the roof, dozens of storeys up, carrying a folded banner. I walked to watch their progress, sickened by the height just looking up at them.

A young American in a Greenpeace boiler suit, part of a group taping off the sidewalk beneath the climbers, addressed me politely. 'Greenpeace demonstration, sir, please can we ask you to keep away from the building until it's over?'

I moved back. Another young Greenpeace man looked at my suit and UN identification badge.

'Are you part of these climate negotiations, sir?' He didn't know me from Adam.

'I am, and I want you to know that I think you people are doing a terrific job.'

Inside the UN the sessions began, only slightly delayed. The secretariat had tabled an entirely new draft text in which they had attempted to find language which might stand a chance of being agreed by consensus. The critical section was on commitments, of course. It had apparently been drafted in Washington the previous week by White House officials and the UK Environment Minister. It was a masterpiece of circumlocution, mentioning the possibility of developed countries stabilizing emissions by the end of the decade, but at an unspecified level, and merely as an 'aim', not as a legally binding commitment. It allowed the Europeans and Japanese to claim that the 'spirit' of the convention as negotiated meant committing – as they all had, unilaterally – to a freeze in carbon dioxide emissions by the year 2000. The Bush

Administration meanwhile could claim to Congress that they had committed to nothing.

The chairman presented the new draft text to the massed diplomats. It was necessary because the old one would have got them nowhere, he said. There were so many square brackets in it that they would have had only three minutes to cover each one.

The Saudis immediately asked for the floor. 'I agree with your assessment that we must speed our work here,' said their head of delegation. 'And we need to be flexible. However, we can't exclude the use of existing text.'

'Yes,' said the chairman, dubiously. 'But remember, please, that we need a convention signed by a maximum number of governments. I say maximum because we must give our governments an effective convention.'

The chairman wouldn't have been this pushy with the Saudis unless he had had informal assurances that the big industrialized countries were all willing to play ball. It looked as though the carbon club might be heading for disappointment.

With just a few days left, and the real chance of a treaty in prospect, however, spectacular brinkmanship was still in evidence. The US State Department proposed a 'rewording' of the tough objective of the convention, as submitted by the EC in Nairobi. The ultimate objective, the USA now said, should mention not stabilizing atmospheric greenhouse-gas concentrations at levels which pose no danger, but the need to control the *rate of increase* in these concentrations from human emissions.

The hall exploded in indignation. Among a string of speakers against the proposal, Japan dismissed it with a brusqueness never before heard from their ultra-diplomatic delegation. AOSIS diplomats audibly choked back their anger at the microphone.

The USA quickly backed off, and tried to turn its concession into a crude quid pro quo. 'I didn't expect to cause such commotion,' said the State Department's lawyer, primly. 'It was meant to be a constructive suggestion, but if there are objections, we will withdraw it.' However, she added, since the USA was being flexible on the proposal about the objective, other delegations could perhaps compromise on other wordings.

Then I witnessed a key moment in history. 'Are there any more interventions?' asked the chairman, hardly looking up from the dais. 'No?' he said, barely pausing for breath. 'Thank you. Article 2 is agreed. Let us now move on.'

The objective, committing governments to the ultimate goal of stabilizing atmospheric concentrations of greenhouse gases, and therefore to deep cuts in emissions, was the first article of the convention to be agreed.

Don Pearlman got up and left the room.

'That was his doing,' Dan Lashof of the Natural Resources Defense Council whispered to me. 'The US delegation were merely showing him they couldn't shoot down the objective.' Lashof, a former US government official himself, had good sources on the delegation. Pearlman, he said, was furious that the convention might now seemingly be agreed by the time of the Earth Summit. Weak as it was, it was too much for Pearlman, Schlaes and their paymasters.

Indeed, I later saw Pearlman giving Dan Reifsnyder the most severe finger-lashing I had ever seen him give anyone. He looked spitting mad. The body language of the two men, fully visible to anyone in the corridor at the time, was very much that of incandescent headmaster and recalcitrant schoolboy. Neither did Bob Reinstein, as head of the US delegation, escape. I later saw him surrounded by Pearlman, Schlaes and two other industry lobbyists. Again, from the body language, Reinstein was on the rack.

On 2 May, the carbon club found a way of dealing with my 'worst-case' opinion survey. The alarming results of the survey, to recap, had shown that almost half the climate scientists surveyed thought it possible that a runaway greenhouse effect could be triggered in a business-as-usual world. More than one in ten (13 per cent, to be exact) even thought it probable.

What happened next was breathtaking in its simplicity and effectiveness. It taught me a lesson I shall never forget. Greenpeace USA's subsequent investigations clarified the sequence of events. Fred Singer's organization, the Science and Environmental Policy Project, took the Greenpeace survey press release and mailed it to selected journalists, along with a covering letter putting their own spin on it. Our press release had been specific about what a runaway greenhouse effect was: it talked about a point of no return beyond which lay unstoppable heating of the atmosphere, a threat to the future of the human species, and indeed to most life forms. Yet somehow, from that point, beginning on 2 May, the survey began to be spun back in the American media as showing that only 13 per cent of scientists thought global warming *itself* was probable.

The process began in the *Washington Times*, where the 2 May editorial raged against what the *Times* saw as the State Department's 'lean to the green' at the last session of negotiations before the Earth Summit. The target was the State Department document 'US Views on Climate Change', which was now proving to be instrumental in the eleventh-hour breakthrough. The State Department, following the IPCC scientists, had merely concluded that – assuming no change in emissions – 'significant changes in the climate system are likely': nothing controversial there, merely an articulation of what the several hundred scientists on the IPCC's science working group had concluded in their reports of 1990 and 1992, and the view that a majority of the world's climate scientists would agree with. But it was lambasted by the *Washington Times*. 'This again is the extremist environmentalist line, and it is flat wrong,' the editorial spat. And here, in the very next sentence, came the evidence. 'Greenpeace, which polled around 400 climate scientists around the turn of the year, was disconcerted to find that only 13 percent subscribe to their alarmist version of the greenhouse warming. Thirty-two per cent said that the theory was possible and 47 percent that it was probably not true.' And where did this leave the State Department? 'Way off to the extreme left,' said the *Times*.

From this point on, the grotesque misrepresentation in the editorial caused a ripple effect. On CNN's *Crier & Co*, Nancy Mitchell of Vice-President Dan Quayle's Council for Competitiveness volunteered in a debate that according to a Greenpeace study, only 13 per cent of scientists polled were concerned about global warming. A Greenpeace USA press officer phoned CNN, but failed to persuade them to correct the statement that Mitchell had made.

An article in the *Wall Street Journal* by the President of the Competitive Enterprise Institute would echo the same line, and he would trot it out again on the MacNeil/Lehrer *News Hour*, concluding that the poll 'suggests there is no consensus that carbon dioxide is a serious threat to the future'. This was exactly the reverse of what it suggested. Greenpeace USA would write to the *Wall Street Journal* complaining about the Competitive Enterprise Institute's distortion, to no avail.

And so it would roll on. Finally, on 27 July, the survey would make the *New York Times*, something the original release – ours – had failed to do. Al Gore's environmental credentials would be the target. Yet another story would confuse runaway warming with IPCC best-estimate warming. 'The survey did not reflect the considerable causes for concern outlined by a succession of international and national scientific panels over the last few years,' concluded the offending article.

The key to the distortion undoubtedly lay with the covering letter supplied with the release when it was sent out by Singer's organization. Jeremy Taylor of the Cado Institute, who had misrepresented the survey on CNBC, confirmed over the phone to a Greenpeace USA press officer that he had received the information from them, along with a covering letter from the executive director of Singer's outfit. Singer's executive director admitted she had sent the Greenpeace press release to 'perhaps ten' journalists.

As a former scientist at the US Department of Transportation, and Professor of Atmospheric Physics at the University of Virginia, Fred Singer surely knew well the distinction between runaway global warming and the type of global warming described by the IPCC. He knew, but he didn't care. His game was disinformation, and he played it well.

By this stage in the negotiations I had been used a number of times as an informal science advisor by Pacific island governments without experienced scientists of their own. I now decided to seek a return of the favour, and asked two island states, Western Samoa and St Lucia, if they would hold a press conference with me. They agreed, and on 6 May the attorney general of Western Samoa, the head of the St Lucia delegation, and I faced the New York UN press corps. The attorney general spoke about cyclones and the insurance problems facing the islands, the St Lucian spoke about rising sea levels, and I spoke about coral bleaching. In this way, we covered the three great threats to the island states, their peoples and their ecosystems.

We faced a barrage of technical and political questions from a full pressroom. Interestingly, no one asked what two sovereign governments were doing holding a press conference with an environmental group. Later, however, a Japanese television director asked the attorney general – an NHK camera pointing at him – 'why are you associating yourselves with a group like Greenpeace?'

'Well,' said the representative of Western Samoa's government, 'nobody else seems to care very much about us here, now do they?'

The Bush Administration, finally, had balked at standing alone in the world community in opposition to a climate convention. A large enough majority of countries had decided to live with the USA's sickening fudge on commitments. And so at the last gasp, on 8 May 1992, just three weeks before the Earth Summit, a Convention on Climate Change was agreed.

After all that had preceded the historic moment, I missed it, having had to return to Europe a day early. My colleagues who stayed told me that when the chairman's gavel came down and the delegates began clapping, the only people with long faces and hands in their laps were the Saudi delegation, the Kuwaiti delegation, Don Pearlman, John Schlaes and the other carbon-club lobbyists.

There was a deep contradiction at the heart of the Convention on Climate Change which would now be taken for signature by heads of state in Rio. The commitments came nowhere close to meeting its objective. But the convention had established a process which required governments to regularly review the science of global warming, the impacts of climate change, and the implementation of efforts to curb greenhouse-gas emissions. This left the door open, in principle, for rapid action to manage the challenge of achieving deep cuts in emissions – by negotiating a protocol or protocols to the convention – should sufficient political will to do so emerge.

What the unreconstructed defenders of the status quo in the Global Climate Council and the Global Climate Coalition feared, along with their backers in the coal, oil and automobile industries, was the fact that the convention represented the thin end of a wedge. If the world at any level accepted the threat of climate change in 1992, Earth Summit year, then the thick end of the wedge in the years ahead could, in principle, be thick indeed. In perpetrating their various abuses of science, the carbon club used money provided, among others, by the American Petroleum Institute, Amoco, Arco, BP, Shell, Texaco, DuPont, Dow Hydrocarbons, the Association of International Automobile Manufacturers, the Motor Vehicle Manufacturers Association, the American Electric Power Service Corporation, the Edison Electric Institute and the National Coal Association. And not just money, we must presume, but also advice. Most of these organizations had seats on the board of the Global Climate Coalition. These companies, and the other more shadowy fuellers of the carbon club who backed the Global Climate Council, had, since the finalization of the IPCC's early warning back in May 1990, sought to keep society locked into what Al Gore called its ten-pack-a-day habit, despite the enormity of the stakes. They had done this knowingly and cynically, with tactics exactly analogous to those deployed by the tobacco industry in the face of inconvenient scientific findings.

And it wasn't as though they were about to give up.

JUNE 1992, RIO DE JANEIRO

From a wooded mountain behind the city, a towering statue of Christ looked down on Rio, arms outstretched as though to welcome the thousands who had assembled for the Earth Summit. But in the streets below, armed police patrolled. Paratroopers lined the road from the city to the conference site; tanks guarded the entrance to the main *favela*. Around the hotels where the politicians, diplomats, glitterati and journalists were staying, the street children were nowhere to be seen. The city had been sanitized for the duration.

The Earth Summit was the biggest media gathering in history. With more than ten thousand press accredited, and more than thirty thousand support people allocated to their activities, there were more journalists in Rio than had covered the Gulf War, more than for an

Olympic Games. Every TV network in the world was there, some of them with as many as four crews. As with the Gulf War, there would be a TV pool run essentially by the American networks. A single pool camera would cover press conferences.

Located at the centre of the sprawling conference complex where the intergovernmental meetings were to be held, the press-conference room and its all-important camera could be booked only by governments. But the big NGOs, industrial and environmental, had ways to sidestep this UN stricture. The Global Climate Coalition didn't even have to share a platform with a government in order to book the facility: the US government simply booked the centre for them in its own name. Greenpeace was to hold an initial press conference jointly with the prime minister of the tiny Pacific island nation of Tuvalu.

John Schlaes took the chair at the Global Climate Coalition's first press conference. As usual, the GCC presented itself as the mouthpiece of business as a whole, but beside the Coalition's executive director sat representatives from Texaco, to speak for oil, and the Edison Electric Institute, to speak for coal. These were the constituencies for which the GCC truly spoke. I had a press pass, a donation to the cause by a Greenpeace-friendly TV producer, and sat at the back of the packed room.

The Coalition seemed to have recovered from their disappointment that the convention had actually been negotiated and were on the offensive again. Their new strategy was to try to switch attention away from the developed countries as the main emitters, and the USA as the number one emitter, by pointing to the growth potential for emissions in developing countries. It was a clever and moderately subtle gambit, and it was to be a major theme of the GCC's efforts at the climate negotiations for a long time after Rio. The idea behind it, environmentalists assumed, was that if sufficient concern could be whipped up about the developing countries, then it might be possible to tie commitments by developed countries to some sort of parallel commitment by developing countries. This, as Schlaes knew, was a concession that developing countries would not make. And why should they? No matter what the future potential for emissions in developing countries, 80 per cent of the greenhouse gases in the atmosphere had been put there by the 20 per cent of us who live in the industrialized world. We had the clear responsibility of leading the way with emission reductions, and indeed the industrialized countries had conceded that at the negotiations.

A few days later, as the journalists filed in for the Greenpeace press conference, the Right Honourable Bikenibeu Paeniu, prime minister of Tuvalu, leaned over to ask me why there were no television crews. I thought I had told his officials, but maybe they hadn't understood, or simply hadn't told him. I explained again that, amazing as it might seem, the one old-fashioned looking camera pointing at us from the middle of the Earth Summit press centre would feed pictures to every television network in the world. In principle, I explained, if these networks wanted to use the pictures, our press conference could be beamed from here into the living rooms of many hundreds of millions of people.

It did seem difficult to grasp, I had to admit. The prime minister had not explored, as I had, the corridors in the television centre, where the TV people worked amid a mêlée of cables and screens in a maze of control rooms. As I had wandered round the control rooms I had seen impatient television news producers tapping their fingers as they watched live pictures of press conferences and speeches, only half-listening to the talking heads, looking for sound bites to put in their news bulletins.

I had already explained to the prime minister about sound bites. I hoped he would be able to produce a few. I knew that he wouldn't be in those millions of homes unless he did.

Tuvalu, formerly the Ellis Islands, was the Pacific neighbour of Kiribati. I had approached the Tuvalu delegation a few days previously, suggesting that their Prime Minister do a joint press conference with us. I explained that Greenpeace had a bevy of press officers in Rio who could whip up interest in such an event, and that he could be virtually guaranteed not to suffer the fate he had at the 1990 World Climate Conference in Geneva. There, he had delivered an emotional address on behalf of his threatened people, but by the time he got to give it there were no journalists left in the press gallery. Prime Minister Paeniu had agreed to the suggestion.

The press conference was full, I now saw. We were benefiting from the fact that Tuvalu was the first island nation to give a press conference. The UN's irritable press chief brought the conference to order, resentful in the extreme that Greenpeace had organized itself such a platform.

'I come to Rio,' Prime Minister Paeniu began, 'to tell you of the fate of my people.' He reminded the scribbling journalists of what climate scientists were now predicting for his homeland, and others like it, unless the burning of oil, coal and gas could be stemmed. Then he described what his people saw happening today. They saw shores being washed away by the sea, a higher frequency of cyclones, prolonged periods of drought. They struggled as it was to build an economy. Now planning had become a nightmare.

The first questioner cut straight to the main point. The climate convention had no targets and timetables for reducing fossil fuel dependence, he said. The USA was in a sense solely responsible for this. How did he feel about President Bush, therefore?

'I am disappointed. There is no more cold war,' Paeniu replied. 'It makes me question the principles of democracy because of his failure.'

'Do you hold him personally responsible?'

'I believe so.' Paeniu considered for a moment, and shrugged his broad shoulders. 'Maybe it is not him. Maybe he is not being advised properly.' The prime minister knew by now all about the carbon club and its activities. 'He has a moral and spiritual obligation to face reality and accord what is right for the people of the world.'

'And if not?'

Paeniu, a deeply religious man, contemplated his next words for a long moment. I could see in my mind's eye the television producers sitting forward in their control cabins.

'There is someone up there who will judge us,' the prime minister said. 'We will all be judged. Even President Bush.'

Prime Minister Paeniu got his message into the world's news bulletins.

Buzzing around the Earth Summit, seemingly with a television camera pointed at him on a permanent basis, Senator Al Gore was in his element. *Earth in the Balance* had recently been published, and was the talk of the town. Environmentalists, hardened to the constant unwillingness of politicians to stick their necks out, no matter how bad the latest news from the ailing atmosphere or oceans, were impressed. I was among them. Even after having talked with Gore several times about climate change, and seeing at first hand how deeply worried he was, I could still not credit that an American politician could write such things.

Gore described climate change as the number one threat to the human future, ahead of all the others that made up what he called the global environmental crisis. He wanted to see the policy response to that crisis become humankind's 'central organizing principle'. As one of his policy recommendations, Gore advocated the phasing out of the internal combustion engine within 25 years. This in a nation where the number one corporation made internal combustion engines and the number two corporation filled them with oil! I could only imagine the apoplexy that his book must have caused General Motors and Exxon.

One passage of Gore's book moved me greatly. 'The voice of caution whispers persuasively in the ear of every politician, often with good reason. But when caution breeds timidity, a good politician listens to other voices. For me, the environmental crisis is the critical case in point: now, every time I pause to consider whether I have gone too far out on a limb, I look at the facts that continue to pour in from around the world and conclude that I have not gone nearly far enough.'

Within months, unknown to everyone in Rio, Al Gore would be on a ticket to the White House with the chance to test-drive those sentiments.

One by one during the week of the Summit, the heads of state signed the Framework Convention on Climate Change. I tried hard to feel good about the TV pictures of the solemn signing ceremonies. Certainly I took some solace in the manifest frustrations of Pearlman, Schlaes and company. They clearly felt they had lost important ground. But I knew that the toothless nature of the treaty meant that it could not provide the stimulus to instigate the fundamental changes needed in energy markets: the beginning of the end for the oil era, and the end of the beginning for the solar era. Legally binding targets and timetables for reducing emissions, no matter how small, would have had a chance of doing that. But not the convention as it stood.

I took time off after Rio. I stayed in Brazil for a month, visiting among other places Manaus in the Amazon Basin. I reviewed the last three years. I felt sure that a new strategy was needed. I was beginning to feel that governments alone would not and could not deliver the goods. It would obviously be vital to maintain progress in the climate talks, and it would be in that forum that victory in the carbon war would ultimately have to be codified. But something else was needed before that day could materialize. A wholly new dynamic.

I and my Greenpeace colleagues had been frustrated for some time that the Global Climate Coalition and their collaborators had been able, unopposed, to present themselves as the voice of business. We knew they were not. They were the voice of a certain category of business: fossil fuels, automobiles and, to a degree, chemicals. They were not the voice of the fishing industry, the agriculture industry, the tourism industry, the water industry, the skiing industry, the medical profession – all the sectors that stood to lose in a world making no effort to reduce the enhanced greenhouse risk. They were certainly not the voice of the financial sector, or most especially of the insurance industry.

I resolved to switch tack after the Earth Summit. My colleague Paul Hohnen had been the first to question why the insurance industry was not represented at the climate talks. Their very solvency seemed to be at stake should the dice roll unkindly in an overheating world. I decided to refocus my efforts, post-Rio, on the financial-services sector. We needed to enlist their help if ever we were to undermine the carbon wall. We needed them to speak out, and better yet take action. Insurers, along with banks and pension funds, were the primary bankrollers of fossil-fuel profligacy, both in debt and equity. There might be a chance that they would come to see this situation as dysfunctional. Why invest in fossil-fuelled energy if burning the stuff is mortgaging your own market? If we could begin to reduce the attractiveness of fossil-fuel investments, and at the same time increase the appeal of clean-energy investments, that might soon amount to a powerful dynamic irrespective of progress at the climate negotiations. Indeed, evidence of proactive engagement by the financial sector would probably *encourage* progress at the climate talks.

In the heart of the Amazon rainforest, I daydreamed about the future flight of capital from carbon to solar, and schemed of ways to make that come about.

4

Blown Cover

July 1992–April 1993

SEPTEMBER 1992, CITY OF LONDON

As I stood outside Lloyd's of London, it seemed impossible to credit that this three-centuries-old institution was in as much trouble as the financial pages said it was. Lloyd's began life as a coffee shop where traders met to arrange insurance for their ships. These days it was a monster of a building in the City of London, resembling with its festoons of pipes some kind of futuristic pollutionless factory, looking out over the headquarters of some of the biggest insurance companies in the world.

The cream of London's financial community now milled around me: the bald and the bouffant, the cropped and the coiffured, their pinstripes expensive and their accents public-school. I wondered how many of them knew about global warming, and the implications it might hold for their recent multibillion-dollar losses, or – more to the point – the prospects it offered for escalating their miseries? I was about to try to find out.

London was the main insurance centre of the world, meaning that premiums flowing into the city from abroad were vital invisible earnings as far as the hard-pressed British balance of payments was concerned. In fact, they were the biggest single source of such earnings. Lloyd's specialized in reinsurance, and one-third of the world's $100 billion per year reinsurance catastrophe premiums flowed through London at the time, half of that through the building I was about to enter. This glass-and-steel monstrosity in Lime Street may not have been at the heart of the British economy, but it was certainly a vital organ.

At the time, the global insurance business took just over $1,000 billion dollars a year in premiums. The insurance industry, I reflected, was bigger than the international arms trade, bigger even than the oil business. Worldwide non-life insurance premiums ran at over $600 billion per year, having doubled in size in less than a decade during the boom years of the 1980s. The business of reinsurance had run fairly consistently at 15 per cent of direct insurance through this period of plenty.

Lloyd's, like most of the rest of the industry, had ended the 1980s badly, and begun the 1990s even worse. A battery of billion-dollar windstorm losses in Europe, North America and Japan had caused a nosedive from underwriting profit to loss – spectacular loss, in the case of Lloyd's. Then, a month before my visit, Hurricane Andrew had made landfall in Florida. Between 1970 and 1992 the insurance industry had taken more than $10 billion in premiums for homeowner and property catastrophe insurance in Florida. Within the few hours Hurricane Andrew took to blast over the state, they had lost all that, and another $6.5 billion besides. The $16.5 billion insurance bill was an all-time record. Had the storm hit downtown Miami, just 30 kilometres further north, the bill would have been well over $50 billion.

Yet within a week of Andrew throwing motor cruisers up the beaches in Florida, Cyclone Iniki had been doing the same in Hawaii. Iniki was the strongest cyclone in more than a hundred years of Hawaiian records. Then Cyclone Omar, also in August 1992, had been the strongest to hit Guam in 16 years. It did not come into the billion-dollar bracket, but it was the third cyclone requiring President Bush to declare a disaster zone on US territory within as many weeks.

This assault by the elements was in the process of ruining many small insurance companies around the world. As for Lloyd's, shortly before my visit the *Observer* had written that 'the time has come when sane people question whether this 304-year-old will survive to the next century, let alone traditionalists' dreams of a fourth century.'

Richard Keeling and David Foreman were typically charming products of the better English private schools, but they had the candour that day of Yorkshire coal miners. 'There is a saying that living memory lasts only five years,' Foreman told me. 'It applies to this industry.' Some of those working in the business had experience of a major European storm in 1976, he explained. But the industry, it seemed, simply tended to forget that such things could happen as time went by. Hurricanes Gilbert and Hugo, and the 1987 and 1990 winter storms in Europe, had caused first nervousness, then alarm, among insurers. For Lloyd's, specializing in the higher-return but higher-risk business of reinsurance, the losses were huge and came at a time when the more astute underwriters had noticed references creeping into the papers to something called global warming, and its product, climate change.

The syndicate for whom the two men worked had at that point done a very sensible thing. They had commissioned a report from one of the UK's leading climate research centres, asking them for the prognosis on global warming. Was there anything in what some climate scientists seemed to be saying about an enhanced greenhouse effect, as of 1991? Were these windstorm catastrophes something they could expect more of? If so, where?

The report duly came in. The climate scientists who had written it strongly suspected that there would be an increase in the frequency and severity of hurricanes in Florida. They also concluded that the syndicate could expect a return of the 1990 storms in Europe. Foreman pushed a copy of the report across the table to me. I flipped through it, and saw familiar diagrams.

His syndicate decided to react to the advice: it limited its exposures in Florida. Then Hurricane Andrew hit. The syndicate saved itself millions. Foreman's syndicate was one of the few in Lloyd's to be growing in 1992, and had shown profits for each of the previous five years.

Significant industry-wide reinsurance price increases resulted. 'We are charging the companies more,' Foreman said, 'one, because we are retaining our exposures, and two, because we are aware now that there is a lot of substance in the effect of global warming on our industry – the return periods of storms are going to be shorter, and the severity of those storms, we believe, is going to be more severe, and therefore the cost more severe.'

Richard Keeling, Foreman's boss, now joined us. Here was a man, as I knew from my researches, massively respected by his colleagues.

In his jolly manner, I could instantly see why. He, like Foreman, was not at all bothered by my tape recorder.

'We are an incredibly backward-looking business,' he affirmed. 'We make our rates by looking backwards. And the problem with looking backwards is that you can walk over the cliff.'

I asked the two of them what they were doing to alert their colleagues in Lloyd's to the kind of conclusions their syndicate had reached about climate change.

'We're trying to stimulate the debate,' Keeling said.

I asked them to peer into the crystal ball. Where was it all leading?

Keeling told me that insurance cover had already been withdrawn in many parts of the Caribbean. He seemed intent on making sure I understood the magnitude of the problem. 'Bear in mind,' he said 'that the Americans, and now the English too, are actually talking about survival.' The problem was, he said, that global warming was still not perceived to be a problem by major chunks of the insurance business around the world.

What would happen to the industry, I asked, if it went right on with business as usual, and the climate scientists turned out to be right about the climate change estimated for the decades ahead?

'Number one,' said Keeling, 'the catastrophic reinsurance pool would start to disappear. Which is happening now. Number two, the insurance support for certain areas would go. This is happening in certain areas now, for example the Virgin Islands.'

And the Pacific Islands, as I knew from my friends in Western Samoa.

'What then happens is you have an uninsured public. You then have political pressure to have governments step in, and they will. There will be pools either backed by government, or governments mandating that the insurance companies provide that coverage. Then after that, if that doesn't work, what do you do?'

His voice tailed off and he looked at Foreman for help.

'The governments take over the full insurance.'

'And what happens,' asked Keeling, 'if the governments then turn round and say they're not sure they're happy about writing those blank cheques?'

'It's taxpayers' money they are losing,' Foreman said.

'It's not losing money so much, it's . . . ,' Keeling looked for the right word, '. . . ruinous. The governments wouldn't mind spending a few billion pounds of taxpayers' money here or there, but it would come to more than that. That's where we get foggy.'

He looked up at me. 'You probably know more about that than we do. I would imagine people actually make hard decisions about which communities you lose.'

He was wrong. I knew next to nothing about this. What I had heard in the last half-hour had not only educated me, it had changed my world-view.

OCTOBER 1992, ZURICH

Swiss Re, the world's second largest reinsurance company, has an elegant headquarters beside the lake in Zurich. I had been granted an interview with one of the company's senior technical analysts. Swiss Re had set up his unit, a special service for risk study of natural catastrophes, ten to twelve years before. There were now twelve people in it – mathematicians, engineers, physicists, meteorologists and geologists.

The analyst took me through the basics, probing my knowledge. I told him I had read a textbook on insurance, and he looked at me doubtfully over his thick-rimmed spectacles, his manner tired.

'The greenhouse effect is long-term as far as insurance is concerned,' he explained. 'The problem in the short term is the proliferation of values, and the increased concentration of values. These at the moment far outweigh the global-warming threat.' He looked at me hard. I could tell he thought I would exaggerate anything he said about the greenhouse issue.

I told him I understood that the industry's problems were multi-faceted, and I didn't want to simplify the problem. But how seriously did his group take the global-warming threat?

'We take the long-term threat very seriously. There is global warming. There will be more water. Sea level will rise. But one knows it only in a general context. That is no use to insurers. The regional effects are not known.' He doodled on his pad of paper, thinking. 'Or at least, only some are known. Hurricanes will have greater frequency, it seems – not necessarily strength. Their path will be able to extend further north, and reach New York, and Tokyo. I don't have to explain that. That's bad enough.'

He pushed a report across the table to me. It was dated April 1992, and was an internal document – a report authored by a Greenhouse Effect Project Team, their latest assessment. He showed me tabulated comments on risk. Most insurance and reinsurance companies, as I had learned, did not have in-house teams of research scientists. Swiss Re was one of the few exceptions. In a pessimistic scenario, I read, the team anticipated an increased risk of flooding from torrential rains, increased danger from storm surges, increasing drought, more bush fires and forest fires, and more thunderstorms, hailstorms and tornadoes. On the other hand, he said levelly, in an optimistic scenario there might be very little change in all these phenomena.

To my surprise, he gave me the report.

He asked me to consider the state of the market, and the scope for interaction of the industry's short-term problems with the longer-term global-warming threat. The retrocession market had collapsed, he told me, referring to a significant-risk, significant-profit aspect of the reinsurance business which had contributed greatly to the miseries at Lloyd's. But still the insurance and reinsurance markets were doing essentially nothing about the problem. Storm rates had increased in England, but not in Europe. Companies were introducing new cover at rates which were simply too low.

He looked out across the lake, selecting his words. 'We are on a train driving in the wrong direction. Catastrophe rates are too low worldwide because people's arithmetic is wrong. They totally underestimate the proliferation effect. The market is expanding, buying in bigger and bigger exposures, at the same time as the retrocession end of the market has collapsed, at the same time as nobody wants to pay a realistic price.' He was warming to his theme now. 'How will the backdrop of global warming hit the market? It is impossible to say. There could be a machine-gun fire of catastrophes. On the other hand, we could be lucky. It could be all quiet on the Western Front. But then things would be worse. Why? Because the machine-gun fire down the road will hit us even worse.'

'If the IPCC scientists are right, then,' I said slowly, 'the entire insurance market is very vulnerable.'

'Oh yes,' he said.

On the plane home from Zurich, I read the Swiss Re Greenhouse Effect Project Team's report, wondering why the company did not make more effort to publicize its concerns. 'The exposure of important concentrations of values such as New York or Tokyo can rise sharply,' read one conclusion, 'with serious consequences for the industry.'

But the Swiss Re team had made no attempt to quantify the risk. How could they? This was the point that had struck me most in both meetings. The industry was no longer entirely walking backwards, looking back into the past. It was casting cautious glances over its shoulder into the future. People like Richard Keeling of Lloyd's, and my interviewee at Swiss Re, knew well that because of global warming the past could no longer be a guide to the future when it came to the use of historical records for the weighing up of hazard and the setting of premiums.

So what could be used as a guide to the future? The climate models were simply inadequate for predicting catastrophe return times in any particular region. Insurers, like climate scientists, were left only with generalities about the global pattern. They were a long way from the specifics they needed at the regional scale for estimating their maximum exposures and setting their premium ratings.

DECEMBER 1992, LA CORUNA, SPAIN

There is more to the risk of oil dependence than the risk associated with the impact of burning it. First, a gauntlet of risk has to be run in delivering it to the marketplace. That risk is difficult to escape, as Exxon found with the *Exxon Valdez*, Amoco had found with the *Amoco Cadiz*, and so on down the long list of multi-thousand-tonne tanker accidents. On the morning of 3 December 1992, a storm along the north coast of Spain duly added the next black mark on the list. While trying to enter the oil terminal at La Coruna, the tanker *Aegean Sea* was blown onto rocks.

The next day, with another storm raging, flying into La Coruna was a white-knuckle affair. I peered down through intermittent cloud as the bucking plane descended. We banked over a spectacular indented coast. Below, I saw narrow bays backed by short sandy beaches, and snatches of pine trees, shaped by Atlantic winds, looking down into rocky coves. Long streaks of matt brown were everywhere in the water.

The morning papers in Madrid had recounted the predictable background to the disaster – the *Aegean Sea* was aged and single-hulled. The seventeen-year-old tanker was carrying more than 80,000 tonnes of crude oil, more than twice the amount spilt in the *Exxon Valdez* disaster.

As we drove into the city, a thick black cloud of smoke hung over the harbour, coming from the headland where the *Aegean Sea* had hit the rocks. Worse was the amount of oil already in the water. The ship had broken in two, and already many kilometres of coast were oiled.

We changed clothes quickly, and set off for a tour. At the first stop, four kilometres from the tanker, we looked across the entrance to La Coruna harbour from the east. The water in the main bay was brown between the angry whitecaps. The stink of evaporating oil was thick and acrid, despite the fresh winds which were keeping a lone salvage tug from reaching the wreck. In the next bay, a smaller, sheltered one, the sea looked as if it was made of oil. Waves crinkled and glooped obscenely onto the shore. The smell made me queasy, and the fumes left me red-eyed.

We sped up the coast ten kilometres from the tanker, to the fishing village of Lorbe. Another tug was trying ineffectually to deploy a floating barrier across the bay. Oil was already overwhelming it. Twenty glum fishermen hung around the dock, their faces being filmed by a French TV crew. They had only recently started receiving compensation for the last oil spill at the La Coruna terminal, we learned. That had been 16 years earlier.

As darkness fell, we headed back to La Coruna. The bar in the

Atlantic Hotel was full of journalists, salvage and oil clean-up contractors, and – a sorry-looking group in a corner – the Filipino crew of the *Aegean Sea*. Listening at the bar, we learned that private contractors from Britain and the Netherlands had come to La Coruna hoping to be given contracts in the clean-up operation. These people had extensive experience in the Gulf, and had equipment crated and ready to be flown in. But the Spanish authorities were taking no offers. To have been seen to need foreign help would have been to make the spill seem more of a problem, prolonging media exposure and boosting compensation claims.

I chatted to the Filipino who seemed to be in charge of the crew. 'This is a delinquent place,' he told me.

JANUARY 1993, LONDON

The year 1993 was not five days old before another oil tanker hit the rocks. This one, 17 years old like the *Aegean Sea*, suffered engine failure in another storm, off the Shetland Islands. By lunchtime that day the flights into Shetland were booked solid with journalists from all over the world. By then, the wreck was the lead story on TV and radio news bulletins throughout Europe and North America. The oil industry had escaped from La Coruna without PR problems and the attendant risk of further regulation, despite all Greenpeace's efforts, but this time, clearly, it wasn't going to be so lucky.

Paul Horsman, Greenpeace's veteran oil campaigner, rushed to appear live on the BBC's lunchtime bulletin, and I made it into the studios of ITN with 30 seconds to spare. I had to tiptoe, puffing, to a chair beside a newscaster with thick make-up on his face. After these live grillings, Horsman spent the afternoon shuttling round TV studios in London, while I gave radio interviews by phone, with calls backing up on the switchboard. TV crews from the USA, Germany, Japan and Canada tramped into the office for interviews.

The common early questions centred on 'Who was to blame? How bad will it be? How much can be cleaned up?'

The captain of the *Braer* had been attempting to pass through the relatively narrow 35-kilometre strait between Shetland and Fair Isle in winds gusting to hurricane force. Water had somehow got into the fuel tank. But Horsman and I sought to put over the message that the blame lay on much broader shoulders than those of a single ship's skipper. It lay with the oil companies for their irresponsibility in shipping oil in ageing tankers (this time it was Ultramar oil). It lay with the government for failing to control an industry so manifestly out of control. It lay with both industry and government for keeping society locked into oil addiction. What was needed, we said – as a starter – was regulation at least as strong as the 1990 Oil Pollution Act which followed in the USA in the wake of the *Exxon Valdez* spill.

'It is time to bring the oil industry under control.' I must have said it fifty times that day. And again, where we could, we dropped in the need to fast-track the technologies that could wean society off oil.

How bad would the spill be? As in Spain barely a month before, twice as much oil as the *Exxon Valdez* had spilt was in danger of escaping from the *Braer*. The tanker had hit the rocks near Sumburgh Head, an area inhabited by rich and diverse communities of seabirds, and by seals and otters. Along the coast were salmon fisheries. Although the tanker had yet to break, storms were forecast. It would be an environmental disaster, come what may, but we would have to wait and see just how big.

How much oil could be cleaned up? Ostensibly, the oil industry had the best it could offer in mechanical and chemical 'clean-up' equipment at BP's Sullom Voe terminal just up the coast. But in the best of conditions, only a fraction could be vacuumed up, and if such seas continued all the clean-up equipment in the world would be irrelevant.

I saw the first pictures of the wreck on a monitor at the London studios of American TV network ABC late that afternoon. They were taken from a Royal Air Force helicopter. The tanker was firm on the rocks, lying slightly to one side, still intact but with brown oil covering the water around her in plumes stretching out of sight up and down the coast.

'Great pictures,' a watching producer said.

'Gotta have them,' cried her male colleague.

The latter showed me into a studio to record an interview.

'Keep it real short,' he told me.

The camera rolled.

'Are you outraged?' he asked.

On day six of the *Braer* oilspill, after almost constant storms, the salvage team had been unable even to begin efforts to drain the remaining oil from the tanker's unruptured tanks. Yet infrared pictures suggested to them that at least half the 84,000 tonnes of oil was still aboard the vessel.

This was when the worst of the weather arrived over Shetland. TV news crews were hurriedly dismantling their satellite dishes as winds of 185 kilometres per hour approached the island in a depression with the lowest pressure ever recorded over the UK. During the catastrophic windstorm of October 1987, the atmospheric pressure in the depression had fallen to 952 millibars. The depression now approaching Shetland measured 915 millibars.

With hurricane-force winds and waves twenty metres high, it was the strongest storm to hit Shetland in a century. I would later hear insurers sound giddy with relief that it had not hit the UK mainland. Scientists instantly speculated on the potential irony here. At the government's Institute of Oceanographic Sciences, oceanographers had some months earlier published data showing a 50 per cent increase in average wave height in the North Atlantic since the 1960s. 'Coupled with the storms of 1987 and 1990,' Institute scientist Dr John Gould now told *New Scientist*, 'there is evidence that we seem to be entering an era of strong storms, which might be associated with global warming.'

Day seventeen of the Shetland incident became day one of the next spill, on Thursday 21 January. Off northern Sumatra, two tankers collided, and one ruptured and caught fire. The *Maersk Navigator* was Danish-owned, and under charter to the Japanese company Idemitsu Kosan. It was carrying nearly 2 million tonnes of oil. The second tanker, under charter to Shell, was empty at the time and still intact after the collision. The fact that the *Maersk Navigator* was only three years old had not helped it, however. At least one of its twelve tanks was ruptured.

The spill was 80 kilometres out at sea, and it would take salvage tugs two days to reach the ship from Singapore. Normally this spill would barely have made a column-inch in a newspaper. Now it made headlines.

The key to success in the oil business has always had more to do with getting the oil to market than with finding and producing the stuff. A hundred years ago, John D. Rockefeller built his infamous Standard Oil monopoly by first establishing a stranglehold on rail transport, and then by seizing control of the early pipelines. Once oil markets moved from the domestic to the international, the struggle for control switched to tankers. Shell emerged as the main challenger to Standard Oil, in large measure because of its early tanker fleet, and the other major oil companies were all operating their own fleets by the time of the First World War. Today, several thousand giant oil tankers, the biggest ships in the world, ply the oceans. They carry at any one time a staggering 500 million barrels of oil – totalling some 68 million tonnes – in holds the size of cathedrals, in ships the size of the Empire State Building.

Before January 1993, this was already a somewhat costly business. With the latest spate of accidents and spills, it held the potential to be vastly more so. The twin problems facing the industry were regulation of shipping, and liability. If the oil industry were forced to pay out huge amounts to increase the safety of ships, and worse, if oil companies were to be made liable for all the costs of their spillages – if the massive inconvenience of the US Oil Pollution Act proliferated internationally, in other words – the costs could even rise to a level approaching unaffordability. Or at least, to a level where all the long-suppressed alternatives to oil suddenly became interesting to the capital markets.

Policymakers in the European Commission had begun work on drafting strict new measures on the movement of oil tankers in European waters as early as day nine of the *Braer* spill. Then came the *Maersk*, and the Danish connection with that ship sent strong ripples through an already agitated European Community. On the day of the *Maersk* spill, the European Parliament voted for a ban on oil tankers more than 15 years old from entering any EC port. Members of the European Parliament also asked member states to set a date for banning tankers without double hulls from EC waters. If such measures were to become EC law, the oil industry would soon be faced with the need to replace virtually the entire tanker fleet. At present-day prices, the bill for that could easily exceed $400 billion.

Of course, those measures were too extreme to stand a chance of becoming law. But how many more spills might it take before these and other measures, equally or even more stringent, reached the statute books?

This was a key question for oil-industry planners and environmentalists alike in early 1993, and it remains so today. For environmentalists, the opportunity to add a huge new bill to the oil industry's operations, so increasing the incentive for solar energy and energy efficiency, had with the Euro-*Valdez* incidents become highly attractive. In a wider context, we had an opportunity for strategic synergy: to add to the cost of carbon when it is shipped – using shipping rules and liability, at the same time as adding to the cost of carbon when it is burned – using carbon taxes and the other policy measures that might well spring from a climate convention with teeth. That was why, in January 1993, so many environmentalists working on global warming switched for the duration to join their toxic-chemicals and wildlife colleagues working on the oilspill issue.

The same will happen next time. And there will be a next time. Shell's former chief J. van Engelshoven summarized the bottom line for his industry in 1990. 'No matter how desirable it is to reach a zero accident goal, we will never get there. Even if we operate on the basis that every accident is preventable, accidents will, unfortunately, still happen.' And when they do, the skirmishes with environmentalists in this particular theatre of the carbon war will resume.

An additional and immensely worrying consideration for the oil-company planner, in the longer term, is that as tankers grow older they are more likely to be wrecked. The boom years for tanker building were in the mid-1970s, before the second oil shock. The fleet is ageing fast. In the early 1980s some two-thirds of tankers were less than 10 years old. By 1993, two-thirds were more than 10 years old, and nearly half the tankers afloat were more than 15 years old. The statistics show that over 80 per cent of marine insurance losses involve ships over 15 years old.

For the oil industry, this has to be very bad news. It was stark in 1993, and the lack of oil spills as visible as *Exxon Valdez* since then in no way means that the problem has gone away. The backdrop is clear. Publics are becoming environmentally sensitized. An oiled seabird means more to a TV viewer these days than one poor dying creature. The image triggers recurrent concerns about the water the viewer drinks, the food he or she eats, and – increasingly – the air he or she breathes. The oil companies know this. They use it in their own advertising. Take BP's unleaded petrol, billed as 'pollution free'. Or ENI's glitzy depiction of its entire oil operation, billed at the Earth Summit as 'the sustainability'.

All this means that politicians are prone to be ever more reactive when a breeze of public concern blows across their desks. What if policymakers in other countries started to emulate the US Oil Pollution Act? What if – horror of horrors – governments began to question the right of oil companies to offload the responsibility for transporting their product onto shipowners?

One answer, of course, would be to build new tankers. But here the oil industry is caught in the vice of its own profit motive. The problem is overcapacity in the tanker market. Minimal improvements in energy efficiency achieved in the wake of the 1979 oil shock, cutting projected demand for oil, were responsible for this. The glut allowed the oil companies to strike very hard bargains with the tanker owners during the 1980s. In consequence, times became tough in the oil-shipping business. Few tanker owners can afford the $125 million it requires to build a new tanker. Indeed, the competitive nature of the shipping market forces tanker owners to cut corners with existing ships. Captains are under pressure to use the most direct routes to market, even if that necessitates navigating through dangerous straits or close to sensitive coasts. Because most of the fleet is registered under 'flags of convenience', whereby countries like Liberia and Panama certify the seaworthiness of both tanker and crew, crews are selected from those nations whose mariners are more likely to accept low wages. Needless to say, accidents are more frequent for flag-of-convenience ships than for those registered in industrialized countries.

The UN's International Maritime Organization does what it can. Its latest requirement is that new tankers should have double hulls, rather than the single hulls that are all that stands between the oil and the ocean in the great majority of tankers today. The IMO also requires tankers older than 25 years to have double hulls. This is essentially a requirement, given the huge costs of refitting, for the oldest tankers to be scrapped.

The oil companies know they have a problem. They inspect the ageing tankers before they charter them. BP's inspection record, detailed in one of the memos routinely leaked to Greenpeace during the *Braer* episode, was instructive. Of more than 3,000 tankers inspected, nearly a third were blacklisted by the company. 'Many of the ships now over 15 years old,' the memo read, 'have not been maintained to the standards we would employ on our own ships.'

In the face of this kind of exposure, the companies set up expert groups to study oilspill clean-up technologies. They beef up oilspill response capabilities. They generally try to look as concerned and responsible as possible. But they cannot escape the inevitable. And meanwhile, the threat not just of PR disasters, but of multibillion-dollar, hyper-liability payouts, is increasing all the time.

FEBRUARY 1993, LONDON

When I returned to Lloyd's of London for the second time, it was to give a seminar. Over the Christmas break I had written a report on my basic case on the implications of global warming for the insurance industry. During January I had the report reviewed within the insurance industry. Lloyd's experts, and the head of the technical team at Munich Re, the world's biggest reinsurer, had ironed out the glitches for me, thus giving the report a vital layer of credibility. Greenpeace offices around the world were set to release the report to the media immediately after my presentation to Lloyd's.

Richard Keeling, recently appointed to the Council of Lloyd's, and chairman for the seminar, met me at the door. 'Don't scare them too much,' he told me with a small grin as a hundred or so underwriters and agents filed into the mock-ancient lecture room in the heart of the ultra-modern building. 'They have so much to worry about at the moment, poor dears.'

Not just them. In Hawaii during December, the Hawaiian Insurance Group – the state's fifth biggest insurer – had announced that its huge losses meant it had to cease trading. Other Hawaiian insurers immediately issued an indefinite moratorium on the writing of new policies. First Insurance, Hawaii's biggest insurer, announced that it would cease renewing existing policies as of 1 February 1993; 38,000 homeowners would be stranded, without insurance for their homes. In Florida things were just as bad. As the full extent of the losses from Hurricane Andrew became clear, amid bankruptcies among the smaller insurance companies, the state insurance commissioner pushed through emergency legislation mandating $500 million in claims to hurricane victims who had been left stranded by the collapsed companies.

Then, on 10 December, New England was hit by one of its worst ever storms. Dubbed 'The Great Nor'easter of 1992' by insurers, it flooded the New York subway, causing it to be closed down for the first time in its history. Hurricane-force winds, heavy rains, river and tidal flooding, and massive snowfalls caused $650 million of insured losses. The barrage dragged houses into the sea, wreaked havoc along 1,000 kilometres of coastline, killed at least 18 people, and caused a state of emergency to be declared in New York, New Jersey and Connecticut. I gave my lecture at Lloyd's to a particularly attentive audience.

I waited to see how the release of the report would go. Greenpeace Germany's ace campaigner Wollo Lohbeck phoned me that night to say that he had six live radio interviews the next morning, and would I please tell him what the report was all about? The following week he clocked up dozens of media interviews. Greenpeace Netherlands and Denmark also reaped superb coverage. Greenpeace Norway opted for an 'exclusive', and this resulted in a long and prominent story in one of Norway's major papers, where the whaling issue normally made good coverage for Greenpeace impossible to come by.

In Japan, Yasuko Matsumoto was over the moon. The newspaper *Asahi Shimbun* gave the report a quarter of a page. The journalist who covered it had done some research of his own on the subject. One of his interviewees, Toshifumi Kitazawa, an executive at the giant Tokyo Marine and Fire insurance company, proved to have – as the article put it – 'the same sense of crisis as his counterparts abroad'. Kitazawa had no doubt what was afoot. 'Behind these events,' he told *Asahi*, 'are the global-scale changes in climate patterns.'

In 1991, in the aftermath of Typhoon Mireille, insurance payouts were ten times higher than had ever been paid for typhoon damage in Japan before. An unnamed source in the Japanese Marine & Fire Insurance Association told *Asahi* that 'if more disasters like [Mireille] follow, it could affect the industry's very existence.'

MARCH 1993, NEW YORK

Governments had agreed to meet beyond Rio for biannual negotiating sessions, to see if there was the collective will to begin giving the convention teeth. The first session of climate negotiations attended by the newly elected President Clinton's appointees began in New York in March 1993. Hopes were high among environmentalists and progressive delegates alike.

I arrived early in the city, in time for the second hundred-year storm in three months. I sat in my hotel room watching on television the progress up the East Coast of the storm of the century. Reporters on location braced themselves against hurricane-force winds as they spoke. Weather forecasters in the studio were unable to conceal their excitement at having to deal with something even worse than the Great Nor'easter of '92. 'We'll stay right where we are until the sea comes to get us,' an old lady on Long Island told the TV cameras. Police were asking such stay-ons for their next-of-kin and dental records. The Governor of New Jersey stood on a sea wall, defending its effectiveness to the camera. Hundreds camped out in closed airports. Reports on Channel Four opened to the ominous notes of Holst's *The Planets*.

A sign on a board across broken windows read 'Oh no! Not Again!'

On the first day of the negotiations, queuing to register in the UN, I asked Don Pearlman if he was not now becoming worried about the suspicious number of freak storms around the world of late. This latest storm was going to cost the economy well over a billion dollars, according to early estimates. In fact, insurance losses alone would eventually reach $1.6 billion, making it the twelfth climate-related billion-dollar disaster for insurers since 1987.

Pearlman snorted. 'I was encouraged by the US Met Service guy on NBC news this morning. He was asked specifically if global warming was to blame, and he said no. Categorically no.'

But the insurance industry was beginning to conclude otherwise, I told him.

'Yeah. That's because they want to hike their rates.'

Also queuing to register as a representative of a non-government organization that day was J. R. Spradley. If ever I needed confirmation that the Bush Administration and the carbon club shared adjacent bedrooms, with an unlocked connecting door, here it was. J. R. had been Bush's man at the Department of Commerce. Now out of government, he gave me his card, which showed him to be an energy consultant. I asked him who he was consulting for, and he introduced me to a stern-looking woman from the US National Coal Association, Constance Holmes.

I wasn't surprised. J. R., a wearer of red braces, snappy suits and the occasional ten-gallon hat, was a personality of some eccentricity, and had been the most enthusiastic of Bush's foot soldiers. He was someone who could be relied on for entertaining company at the dinner table, but his ideology perfectly encapsulated the complete antithesis of my own. When called upon to do so, the other Bush climate-change hit-men would be content with offering as robust a defence of the administration's line as they could muster. Some would even allude, in private, to discomfort with their task. But not J. R. Spradley. He was an advocate – a missionary, even. He wanted to convert, to explain. He wanted to take his argument to you. J. R. was big on adaptation. He had once created a storm among island delegations by asking famously 'what's wrong with a bit of sea-level rise? It is merely changing land use – where there were cows there will be fish.'

J. R. now gave me a fat report on how a carbon tax would be bad for the US economy. A carbon tax was one of the main potential vehicles for cutting greenhouse-gas emissions. By taxing fossil fuels, and not renewables, the US government could force a certain amount of switching away from dependence on oil and coal in the economy. Most European Union countries by this time favoured such a tax. All eyes were on the Clinton administration. If they could push a carbon tax through, the remaining foot-draggers in Europe – notably the UK – could perhaps be coerced into going along with the measure.

President Clinton was offering environmentalists hope. 'The idea of a tax shift to discourage polluting is right in line with the principle of making polluters pay,' he had said earlier. 'I would consider supporting such a revenue-neutral tax if it could be accomplished without unduly hampering our industrial competitiveness or raising utility consumer rates.'

The game was on. It was to be the key policy debate of the months to come. If some form of carbon-based energy tax could be forced through in the USA and Europe, the world community would be on the way to reducing greenhouse-gas emissions. The industry lobby groups were fighting this prospect with everything they could lay their hands on. They saw it as another potential nail in the coffin of oil and coal – and they were right.

In the sessions, everyone waited to hear what the first speech by the Clinton–Gore team would contain. Would they commit to cuts in emissions? Would they say they were going for an energy tax? Would their actions match the rhetoric of the campaign trail?

The speech was given by Madeleine Albright, the new US Permanent Representative at the UN. Among the listening hundreds, I saw J. R. Spradley sitting with Don Pearlman.

Climate change was a great threat to the integrity of the global environment, Ms Albright told the massed diplomats. The USA would review all its options to reduce greenhouse-gas emissions. By June, it would have made significant progress with that review. The Clinton administration aimed to have a new plan available by August.

There were no pledges, no signs of urgency. Environmentalists were uniformly disappointed, but knew that with hard work they would have a chance to influence events during the spring and summer.

APRIL 1993, ORLANDO, FLORIDA

At the US National Hurricane Conference, held in a hotel close to Disney World, I listened to the General Manager of one of the Caribbean's biggest insurance brokers, General and Marine, paint a grim picture of the region's full-scale insurance crisis. Finding enough capacity for his clients' insurance needs had suddenly become a nightmare, he said. 'If anyone had told the reinsurers ten years ago that they could face $20 billion in insured catastrophic losses in one year, they would have been laughed at.' This was a reference to the record 1992 global total. In consequence, coverage was either being withdrawn, or rates and deductibles were being hiked to unaffordable levels. This was now one of the biggest threats to the Caribbean economies, he said. Problems even extended to foreign exchange, since 70 per cent of the average premium went out of the country.

How were existing policyholders reacting? Cancelling or reducing coverage in many cases: leaving themselves in the hands of God, in other words. This in turn was leading to problems for banks and other financial institutions, who were quickly realizing the extent of their potential exposure from the very high hurricane deductibles and the fact that some of their mortgagees were cancelling or reducing coverage.

Florida was now facing the kinds of problems that had beset the Caribbean during the previous 18 months. Over thirty insurance companies had by this time tried to quit the state. The Florida insurance commissioner had slapped a moratorium on them, requiring them to stay put for the duration of hurricane season. In Hawaii, it was even worse, *Business Week* reporting that rating agency A. M. Best had estimated that 40 per cent of the population might be without home insurance by the end of the year.

I thought of it as the climatic domino effect. First the insurers felt obliged to withdraw from Pacific and Caribbean island nations. Then, after Hurricane Andrew and Cyclone Iniki, they had felt obliged to try to withdraw from Hawaii and Florida. Then would come coastal areas more generally. And so on.

It had been Hurricane Andrew that had started the panic in the industry, and the stories I heard in Orlando that week showed why. 'We were only a fraction away from a tremendous amount of civil disobedience,' said the mayor of Homestead, one of the two cities affected. 'Had the hurricane hit Miami or Fort Lauderdale, the result would have made the LA riots look like a picnic.' Speakers from the emergency services spoke of the horrors of 'ground zero'. Their films, and countless slides, showed scenes of carnage: a marina where more boats sat in trees and on buildings than in the water; an airfield with planes smashed into tangled piles in their hangers. And literal mountains of garbage. The garbage created by the hurricane in South Florida amounted to 22 million cubic metres: some 22 years' worth of landfill, created in three hours.

'As we go into the 1993 hurricane season, I feel like I'm staring down the barrel of a loaded shotgun,' said the director of Dade County's Office of Emergency Management.

Among climatologists at the time, as now, opinions differed as to whether hurricanes would become more numerous and powerful in a warming world. Given the uncertainties, few were prepared to stick their necks out, and this was understandable. All things being equal, warming waters would extend the areas in which tropical cyclones could be generated, and would probably spawn – and maintain – more energetic storms. But all things would not be equal. For example, in the Atlantic, cooler northern waters seemed to mean fewer hurricanes, and the North Atlantic was one of the few places on earth suffering an overall cooling. Then there was the El Niño phenomenon. El Niño years heralded fewer hurricanes in the Atlantic, and some meteorologists suspected that global warming might generate more El Niños.

It remains difficult to be sure that a warming world will – or will not – suffer more hurricanes, or stronger hurricanes. But this absence of certainty is a long, long way from an absence of risk. And when one looks at extreme events as a whole – factoring in droughts, wildfires, floods, hail, and the rest – it seems certain that the numbers of catastrophes will continue to rise. Indeed, growing numbers of scientists were by this time suspecting that the overall pattern of catastrophes was already showing a global-warming signal.

Munich Re had meanwhile called on governments, businesses and insurers alike to take immediate action in the face of what they were now calling the dramatic development of natural catastrophes. 'The threatened climatic changes demand urgent and drastic measures,' the company's press release said.

Both Munich Re and Swiss Re were now telling the press that the recent pattern of losses must at least in part reflect enhancement of the greenhouse effect. *Lloyd's List* published a long commentary on the analyses of both companies, concluding that 'the convenient theory that the increase in the size of losses is mainly a reflection of higher wealth – and consequently, of insured values – in those countries affected by natural disasters seems to be incorrect. It is far more likely that other causes, such as climatic changes, have already taken over as the main factors pushing losses upwards.'

I felt confident, in the spring of 1993, that all this evidence of brewing economic harm from the early signs of global warming would allow Bill Clinton and Al Gore to deliver on the rhetoric of the campaign trail, and the famous book, when the new US position was announced in August. With the government of Tuvalu, I had devised what I hoped would be an effective campaign to assist with the process.

5

Burned by Warming

May 1993–March 1994

MAY 1993, HOUSTON

Cognitive dissonance: an uncomfortable awareness of the distance between your own view of something, and the views of the people in whose company you find yourself. It's the kind of feeling that makes you ask yourself whether you can possibly be right, or even sane, when so many others seem so sure you are wrong. No forum could possibly offer an environmentalist a bigger dose of cognitive dissonance than the Offshore Technology Conference. Thirty-five thousand people registered for the 1993 OTC in the Houston Astrohall. I was among them, invited by the organizers to take part in a televised debate.

The event offered the promise of witnessing serious discord, for the first time, in the hydrocarbon family. The organizers of the debate intended among other things to pit petroleum-industry advocates of oil and gas against one another. I was along to provide a third dimension: opposition to both oil and gas.

The *Oil and Gas Journal* had flagged this emerging division in the hydrocarbon world at the beginning of the year. 'Industry must avoid gas versus oil civil war,' the title of an editorial had yelled. 'The potential exists. The incoming Clinton administration seems not just to like natural gas but also to dislike oil.' Global warming, that environmental scare story, was at the root of the problem. Clinton's people seemed to be shaping up to exploit the fact that gas produced less carbon dioxide when burned than oil does, and much less than coal. It would all come to no good unless the gas industry stood side by side with the oil industry.

I walked into the Houston Astrohall to find an indoor oil-industry toyshop the size of a small town. Two of the latest helicopters hung from the ceiling. Portions of drill rigs leapt from the floor. Gleaming pipes and drill bits marched in ranks. Valves, blow-out preventers, drilling-mud additives – all the paraphernalia of the industry arrayed in stall after stall along aisles stretching away at all angles into the towering interior of the building. Groups of suited figures studied the forest of wares. I took a closer look at the contents of a few stalls myself. I could feel the faint pull of my past.

My companions on the TV debate panel were Daniel Yergin, famous oil-industry analyst and author of *The Prize*, the Canadian Energy Minister, the editor of the *Middle East Economic Survey*, a vice-president of a major gas company – Coastal Corporation – and a DRI McGraw-Hill oil economist. The economist, I knew already, was a creature of the carbon club, much used by the Global Climate Coalition. The lights went down, and I revisited the ground I had first covered almost two years before in The Hague: defending the Greenpeace cause in front of an oil-industry audience measured in hundreds. This time there were two differences. My opposition was split. And a television camera was rolling.

I watched with growing delight as the debate unfolded. The man from the gas company agreed with my environmental threat assessment. Yes, he said, climate change was a major problem, and we had to do something about it. The most trenchant of the oil advocates turned out to be the economist. How, he wanted to know, would it ever be possible to turn economies away from oil? The rhetorical exchanges between the two reminded me in their vehemence of debates between environmentalists and industry spokespeople.

At one point the gas man turned to me. He couldn't understand, he said, why I didn't join him in arguing for gas and against oil. Surely I accepted that gas had to be a major part of the answer to global warming?

The problem was in the carbon arithmetic, I told him. Stabilizing atmospheric greenhouse-gas concentrations required deep cuts in emissions from all fossil fuels. We had to get to a solar future running on renewable energy sources and energy efficiency as soon as possible, and right now the expanding gas industry was busy setting itself up to make things worse, not better. It was taking money away from renewables and efficiency. Given that we couldn't turn to solar power overnight, gas might have to have a role as a bridging fuel to a future running on renewables, but we had to be very aware that there was a vast amount of gas available. There was around 1,000 billion tonnes of carbon in gas below ground, according to the best estimates then, compared with some 200 billion tonnes in oil. You couldn't allow the oil industry to reinvent itself as a gas industry, especially not in the face of arithmetic like that: 300 (maybe 200) billion tonnes of carbon from fossil fuels of all kinds was probably more than enough to risk catastrophic destabilization of the climate. And on top of that came the problem of leakage. Natural gas was, after all, primarily methane. Just a 3 per cent leakage of gas from the production, transportation, distribution and use of gas and you would lose the advantages of its lower carbon intensity with respect to oil. At twice that, you were not even beating coal.

After the debate, a long queue of industry people hung around to talk to me. Most were on the younger side. The tone was overwhelmingly supportive: Keep going. Don't give up. You're not alone. There are people in the industry who share these views.

One man took his turn. 'Jeremy, you don't recognize me!' I didn't, notwithstanding the English accent. He spoke the name of a former classmate from when I was 15 or 16, in Hastings, Sussex. We went to the bar. He told me how he had come from Hastings, England, by stages to Houston, Texas. He was now international marketing director for Schlumberger, the oil industry's biggest supplier of well-site monitoring equipment.

He had missed most of the debate, and made me replay my case. I did so in outline, though with no hope that someone of his seniority in an oil-industry service company would be in a position to sympathize.

But I was to be surprised, once again. Schlumberger wouldn't necessarily have a problem with a future run on renewables, he said. The company was into monitoring, instrumentation, that sort of thing. They didn't necessarily have to stay locked into oil and gas. The instrumentation of water supply, for example, was going to be a big deal for them in the future, global warming or no global warming.

And in my renewable-energy scenario, he said, he could see a great market getting going, say, in the instrumentation of offshore wind farms.

I bought him another beer.

MAY 1993, BARCELONA

I speak mostly of oil, it seems, and just latterly in the story, of gas. But what of the third and biggest brother in the fossil-fuel family? Oil may be the biggest contributor of carbon dioxide among the fossil fuels burned today, but in terms of carbon reservoirs underground, and hence the amounts that can be burned tomorrow, it is the smallest. Coal is king in this regard. With the quantities left below ground measured in thousands of billions of tonnes, we could in principle still be burning large amounts of coal several hundred years from now.

So the coal industry's foot soldiers fight even harder than those of the oil industry in the trenches of the carbon war. All the big national and international institutions provide troops. Prominent among them are the US National Coal Association and the World Coal Institute. Among the corporations and institutions funding the Global Climate Coalition, there are just as many from the coal industry as from the oil industry. In fact, their collective process of denial is even bigger. The oil industry is generally willing to talk with environmentalists, and tends to do so civilly. On the whole – in my experience – the coal industry treats environmentalists like plague carriers, and when discourse does takes place it tends to descend swiftly into the gutter.

From Houston and the oil-and-gas industry's biggest junket, I flew on to Spain for a rare dose of dialogue with Big Coal. I was gratified to find the mood at the Sixth US–European Coal Conference immediately and tangibly beleaguered. A high-ranking official from the European Commission gave the assembled coal executives a grim opening survey of the European coal market. 'I hope coal will keep an important place in the market, but as you know there are clouds on the horizon. The most important one is environmental.' The role of environmental factors would only increase, he emphasized, as the Commission was forced to react to public pressure.

A procession of speakers perfectly reflected the plunging morale of an industry which had thought in the early 1980s that steam coal would reach $100 a tonne, and were now lucky if it topped $10 a tonne. A banker from Lehman Brothers gave an overview of the US market. 'Coal demand and spot prices in the US have been weak due to a prolonged period of difficult economic conditions and unfavourable weather trends,' he began. He presented statistical data showing how the USA was getting warmer. 'Many weather forecasters say the weather goes in cycles,' he suggested. 'Maybe there will be a return to cooler years. I don't know.'

This all sounded a bit familiar.

Even at sea, the news was gloomy. Given its problems in Europe and the USA, the coal industry was banking on expanding markets in Asia, and would have to make much greater use of the bulk-carrier coal trade if Malaysia, Indonesia, Thailand and all the others could be persuaded to buy American and European coal. But this would be far from easy, the Director of P&O Bulk Carriers told the conference. The cost now for shipping coal across the Atlantic was around 50 cents a tonne. And the bulk-carrier fleet was ageing: close to 25 per cent of the ships were 16 to 19 years old. In the last three years, 13 had sunk.

This sounded familiar too.

The time for the environmental session arrived. In the chair was Constance Holmes, the chair of the Global Climate Coalition's international committee. Holmes introduced the first speaker, Dr Harlan Watson, a former member of the Bush Administration's team at the climate talks. Watson was a dapper man with a dramatic style, which he used to good effect. He began with a straight factual account of the history of the climate negotiations. He built up to his main message without excess rhetoric, but when he got there the message was hard.

'Dire warnings of doom and gloom and apocalyptic forecasts are the standard fare of nearly any discussion of the climate change issue. If I were to follow that time-honoured tradition, mine would be the following: the global climate change issue and the UN Framework

Convention on Climate Change threaten the coal industry worldwide – particularly in the industrialized countries – with extinction. I am not prepared, however, to go that far. I can assure you, however, that you need to take this matter very, very seriously.'

He looked around to see how this was going down. 'Industry, with few notable exceptions, has failed to appropriately engage the issue. In fact these exceptions are so notable, I want to digress a moment to recognize three individuals.'

I was about to hear further evidence of how things had been between the Bush Administration and the carbon club's hit men. Harlan Watson had been on the committee in the White House's Office of Science and Technology Policy, dealing with US global climate-change research. For four years he had been scientific advisor to Bush's secretary of the interior.

'Don Pearlman, executive director of the World Climate Council; Connie Holmes, senior vice-president for policy of the National Coal Association; and John Schlaes, executive director of the Global Climate Coalition. As a matter of fact, Don is so dedicated that he declined to appear at this session because it conflicted with an IPCC meeting in Montreal.'

Where, I would later discover, he had been up to his usual tricks with more than usual effectiveness.

Watson reached his recommendations. 'What should you do, you might ask? Let me make several suggestions. First and foremost, you must put aside your differences and get properly organized to address the issue. You need to speak with one voice. Second, you must get timely, credible and relevant information to the political decision-makers, to the media and to the public at large. Third, you need to follow closely the activities of both the INC and the IPCC, and, as far as possible, actively participate as NGOs through trade associations.' Then, finally, to the bottom line. 'Do not underestimate what you are up against. In the US, it is the combined forces of the environment community and Vice-President Gore and his powerful allies in the Administration. In the past, business interests throughout the world could rely on the United States to maintain sanity in the international environmental arena – this was certainly true during the Climate Change Convention process. Well, my friends, that is not the case today, and it is time to pull up your socks, roll up your sleeves, and get to work.'

Watson finished, relishing the applause.

Constance Holmes fixed me with her hard stare. It seemed to be my turn. The subsequent scene was best described by the *Energy Daily*, an industry journal, a week later: 'Global warming had an apparent cooling effect at the sixth US–European coal conference last week, as industry representatives emptied out of the conference room before an address on the issue by an internationally known environmental representative. As Greenpeace's Jeremy Leggett took the stage, attendees left in droves, leaving perhaps 30 coal-industry representatives to listen to their opposition.'

After the session was over, they took me off to another room, and a press conference for the benefit of the half a dozen journalists covering the conference. Most of them worked for coal-industry journals. Constance Holmes again took the chair, and launched straight into the question she evidently thought would most effectively skewer me. How could the West provide the technologies necessary to reduce carbon dioxide emissions in the Third World if revenues were being drained by carbon taxes?

Just watch and see how revenues will be drained if global warming takes off as the IPCC predicts, I said. And anyhow, who said carbon taxes would have a negative economic impact? The carbon-fuel industries, mostly. Many studies suggested exactly the reverse, and the tax could even be made fiscally neutral.

Harlan Watson at that point volunteered to give the trade press his view of what was motivating environmentalists in their advocacy of carbon taxes, and indeed any measures to limit greenhouse-gas emissions. The problem, he said, was that after the collapse of communism, and the exposure of command economies as recipes only for disaster, the old Left had found themselves with nowhere to go. They had elected in large numbers to switch to the environmentalists' bandwagon.

Fighting hard to keep the flame away from the blue touch paper now, I tried to paint a picture of a representative group of my colleagues for the journalists, and show just how many million miles Watson was from the truth. I invited them to consider the head of Greenpeace's delegation at the climate talks, the senior diplomat who would by now doubtless have been an ambassador; the concerned lawyers working on a fraction of the salary they could have made in industry; the intelligent young graduates who applied in hundreds for every grinding administrative assistant's job.

After we finished answering questions, I turned to Watson immediately. I had managed to put the flame out by now. 'You were in danger of getting a bit near to your philosophical underpinnings there, Harlan,' I laughed. 'Did I understand you correctly? Do you seriously think that we are all old communists?'

His shiny face was twitching as he tried to hold my gaze. 'I think there are many who have that agenda,' he said in a constricted voice.

JUNE 1993, FLORIDA CITY / WASHINGTON, DC

On the morning of Friday 11 June 1993 I stood in the corridor of the Willard Hotel, Washington, with four Secret Service agents, waiting for the prime minister of Tuvalu to emerge from his suite. I knew the agents quite well by this stage. I had watched them guard the prime minister for nearly two weeks. I chatted with Casey Jones, who had been one of President Reagan's favourite agents. Casey Jones wasn't his real name, of course, just the name on the card he had given me in Florida, on the first day of the trip. Casey and I had become easy acquaintances, him with his shoulder holster and me with my sheaves of papers. The door opened, and the cheerful face of the Right Honourable Bikenibeu Paeniu appeared. As ever, Prime Minister Paeniu was in good spirits. Jones spoke into the tiny microphone on his lapel to alert the agents on the street outside, and we set off for the White House.

Enele Sopoaga, Tuvalu's secretary for foreign affairs and the PM's aide for the trip, fell in beside me. 'Did you see the paper this morning, Jeremy?' he asked.

I had. The news was dreadful. A front-page headline in the *Washington Post* had announced that Bill Clinton had abandoned the carbon tax. Only two things in life are certain, Benjamin Franklin once said, death and taxes. But not taxes on energy or pollution, it seemed.

Sopoaga had been told all along by the State Department that neither Clinton nor Gore could see Paeniu, despite entreaties from the Tuvaluans that they wanted only a few minutes to register their case and leave a detailed letter. The most he could hope for, they had told him, was a meeting with the president's environment advisor. Yet now, at the last minute, the Secret Service men had told Sopoaga that the meeting was to be in the White House itself.

Our strategy hinged on the gamble that Prime Minister Paeniu would be given some form of audience by Bill Clinton or Al Gore while he was in Washington. Paeniu had written letters to the leaders of the richest industrialized countries, the Group of Seven, asking them to undertake emergency action on climate change at their forthcoming G7 summit meeting in Tokyo, a few weeks hence. He had told the press that he hoped to deliver his letter to President Clinton in person. If the president and vice-president ignored him, so would the American media, and so too, almost certainly, would the Japanese and the rest of the G7 heads of state.

The case Prime Minister Paeniu brought to the G7 leaders was simple. Tuvalu was beginning to experience the first signs of the sea-level rise and climatic destabilization that would make it uninhabitable unless greenhouse-gas emissions were cut. The G7 nations were letting their Earth Summit commitments slip just when it was becoming clear, in fact, that what was needed were actions on climate change much stronger than those agreed in Rio.

The energy tax issue meant that Paeniu's tour of the USA could not have been better timed. The Clinton Administration had opted for a BTU tax – a tax on the thermal content of fuels (as measured in so-called British Thermal Units). This would have the effect of discriminating against oil, thereby promoting a reduction in oil imports, and making coal and gas sufficiently more expensive that energy efficiency would be stimulated. Renewables such as solar and wind power were to be exempted from the tax.

Clinton had been experiencing great difficulties pushing the tax through Congress at the time Paeniu arrived in the States. The carbon club was up in arms. The oil companies hated it, the coal industry hated it, the heavy electricity users hated it. Their lobbyists were out in force. In the Senate, things had looked bad from the start. Clinton's defence of the tax was almost entirely fiscal – emphasizing the need to raise revenues to help cut the federal budget deficit and to bankroll federal expenditure. Despite his strong rhetoric about climate change during the campaign, the president had chosen not to stress the clear environmental imperative for the tax.

And so we found ourselves, on the morning of Friday 11 June 1993, climbing into a five-car motorcade to travel the few blocks from the Willard Hotel to the White House. The prime minister and Enele Sopoaga rode in a big black limousine, as they had throughout the trip. I travelled in a smaller limousine at the back of the motorcade. All of us were wondering, as the White House swept into view, who would be there to greet us.

The trip had begun in Florida on 1 June, the first day of the 1993 hurricane season. The prime minister's intention had been to express solidarity with fellow cyclone-threatened communities, and hopefully in the process register his presence on the Washington radar screen. I was to be with him throughout, as his technical advisor.

We rode south from Miami in a Secret Service motorcade to the area where Hurricane Andrew had struck. In Florida City, a mobile home had served as an impromptu city hall ever since the hurricane. All around, the humble residences looked like they had sustained an artillery barrage – and days, not months, before. Many were still roofless and boarded up, insurance policy numbers daubed on their walls. The prime minister and the local mayor agreed to enter into a 'sister city' arrangement. They explored the similarities between their communities. Both populations were 10,000; both had a very low per capita income. And as if the risk of more intense cyclones wasn't enough, both communities were within a few metres of sea level, and drew their fresh water from shallow aquifers susceptible to salinization.

There were also differences. Tuvalu's nine atolls sit among more than a million square kilometres of Pacific Ocean, a total land area of 25 square kilometres in an area of ocean the size of France and Germany combined. The crime rate in the island state was such that there was rarely more than one person in jail. Not so Florida City, where the misery and hopelessness following the hurricane had compounded the existing social problems common to many low-income American communities.

An Associated Press photographer positioned the two men in the sweltering sun in front of collapsed houses and piles of debris. Prime Minister Paeniu, it emerged – incredibly – was the first dignitary to visit Florida City since the hurricane. The others had all gone to the bigger city of Homestead.

Driving back to Miami, I fell into conversation with the two Secret Service men I was sharing a car with. One of them was locally based. We passed an ice-manufacturing plant on the outskirts of Coconut Grove. 'There was a riot here after the hurricane,' he told me. 'People were just cut-throat.' He shook his head. 'It was kind of surreal. Like a science fiction movie of the end of the world.'

Soon after we arrived in Washington, Enele Sopoaga and I went to the State Department to plead for a presidential or vice-presidential audience for the prime minister. The new head of the US delegation to the climate negotiations greeted us. No longer would I be dealing with Bob Reinstein in this position. The new man was Rafe Pomerance, formerly ace climate lobbyist for the World Resources Institute, and long-time collaborator with Paul Hohnen and myself at the negotiations. For the first two years of the climate negotiations, Pomerance had been the scourge of Reinstein, and George Bush's other hatchet men, J. R. Spradley and Dan Reifsnyder. Now Reinstein had gone the way of all senior Bush appointees, Spradley was working for the carbon club. And Reifsnyder? He was sitting to the left of Pomerance, who was now his boss.

The president had committed to put a cap on carbon dioxide emissions in the year 2000, Rafe Pomerance reminded us in his opening remarks. That was much more than the previous president had done. President Clinton had, rather courageously, Pomerance thought, proposed a BTU tax. 'In fact, today, er . . . that would probably be the single biggest measure we could take to limit greenhouse-gas emissions. It is a shame that Congress is so resistant . . .'

He tailed off. I had sat in countless joint strategy sessions with Rafe

Pomerance at the climate negotiations. I knew that in his heart he didn't see the threat of climate change any less seriously than I did.

'We know and appreciate how sensitive these matters are,' Enele Sopoaga said. 'But you have to appreciate how desperate things look from our point of view.' Cyclone Nina had wiped out 165 homes with one wave, he said, yet there was clearly far worse to come unless greenhouse-gas emissions were cut, not merely capped, and atmospheric concentrations ultimately stabilized, as the objective of the convention required. Tuvalu was depending on American goodwill and leadership to make this happen.

Pomerance looked thoughtful for a moment. When he spoke, his tone was tentative. 'You know, there're scientists that say that even if atmospheric concentrations are stabilized, the sea level will continue to rise. When you think of it that way . . . it gives you, er, a whole new angle on the problem.'

He tailed off again, without elaborating.

I looked up from my note-taking. 'If I may, Mr Secretary,' I said to Sopoaga, who nodded. 'Of course there's a lag, Rafe. But the hope for the island nations has to be that the climate sensitivity proves to be low, and deep cuts are achieved as soon as possible. You know that. Doesn't that make a case for accelerating emissions reductions?'

At this point Dan Reifsnyder chipped in. 'In Tuvalu, are people looking for action from the big emitters, or looking at what they can do to adapt?'

I looked at Pomerance, who was studying the carpet.

'This is about treatment,' Sopoaga said, a faint edge in his voice. 'We know how busy the president and vice-president are, but we feel that this issue must be important enough to warrant a small amount of their time.'

Pomerance looked embarrassed. 'You have to realize,' he said, 'that this is, er, a very intensely focused moment for the president. All his attention is on the Congress.'

He seemed lost for the next sentence. Another of the officials came to his rescue. Out of nowhere, he elected to give us an account of the good work the United States was doing in coastal management studies.

The meeting finished. Assistant Secretary Tim Wirth, Pomerance's boss, was supposed to have been next on the schedule, but he was running late, an official told Sopoaga. Instead, we would be seeing Ken Quinn, deputy assistant secretary in the Asia–Pacific bureau. Quinn, ex-National Security Council and chief negotiator with the Vietnamese on the missing prisoners of war issue, proved to be a bluff man who was clearly not about to be embarrassed by a nation the size of a single Washington suburb. Many, many heads of state had tried to meet the president, he said. But it had been impossible to arrange.

I risked an intervention. I couldn't understand why the White House didn't want to use this visit to help themselves in presenting the economic package, I told him. The energy tax was a centrepiece of that package, and they could use Tuvalu's plight to take the moral high ground when defending the tax. Surely Americans wouldn't want to think that all those thousands of marines who died to save Pacific islands fifty years ago sacrificed themselves in vain? Surely, at least, they could afford a photo opportunity?

Quinn leant forward. 'Let's be clear, you're talking about a photo opportunity, not a meeting? That's two very different things, you know.'

'We are talking about a meeting, but a photo opportunity if that cannot be arranged,' Sopoaga said.

Quinn told us he would see what he could do. But he wasn't making any promises, we had to understand.

And so, two days later, our motorcade swept round the road south of the White House. I looked across the lawn at the famous building. Behind one of those windows, a few nights before, Bill and Hillary Clinton had watched *Jurassic Park*. An hour and a half devoted to Spielberg's fantasies. Would the president devote a few minutes to Tuvalu's real-life crisis? As for the vice-president, I thought of the words in *Earth in the Balance* which had once so encouraged me: 'now, every time I pause to consider whether I have gone too far out on a limb, I look at the facts that continue to pour in from around the world and conclude that I have not gone nearly far enough.'

The cars pulled up in line at a gate leading to a driveway between the White House and the Executive Office Building, where the president's environment advisor worked. The Secret Service had radioed ahead, and we were waved straight through. I held my breath as we moved up the drive. Would we go right, into the White House proper? Or left, into the Executive Office Building?

We turned left.

JULY 1993, TOKYO

Prime Minister Paeniu's trip to America had been a disappointment from beginning to end. The core strategy, or perhaps more exactly gamble, had failed catastrophically. I told him that on no account would I or my organization withdraw our support if he made the decision to go to Japan anyway, but that we would quite understand if he opted now to cancel that leg of the trip, since our chances of registering his message were probably close to zero.

No, the prime minister told me firmly. He had told everyone we would be going to Japan. We were going to Japan.

The Japanese government could not have been more unwelcoming. The Japanese foreign and environment ministers would be unable to see Prime Minister Paeniu, they informed Tuvalu's Ministry of Foreign Affairs. Neither would any of their officials. He would be given no security, and no form of official welcome.

But the first evening we were in Japan, a strange thing happened. A succession of television stations from Japan and Australia visited our hotel to interview Paeniu. Next morning, he had interviews scheduled from first light, the first with *Asahi Shimbun*, with a readership of 8 million, Japan's most influential newspaper. Yasuko Matsumoto and her Greenpeace Japan colleagues had somehow succeeded in interesting the media in Tuvalu's mission. In the afternoon they had scheduled two press conferences, one for the Japanese press and one for the international press, and these were already filling up with acceptances. It seemed to me that my colleagues had pulled off a miracle.

The NHK evening news showed a long feature on both its main bulletins, to a total audience of around 15 million. The next morning, *Asahi Shimbun* ran a major article. Yasuko translated it for me excitedly over breakfast. 'If global warming continues, island nations are in danger of disappearing,' the headline read. Prime Minister Paeniu had spoken to the journalists of 'genocide by environmental destruction' should global warming be allowed to go ahead unmitigated. German TV, and a wide range of representatives from the Japanese and international press, had turned up to the press conferences. Around the world, others were picking up the pictures and stories from Reuters and the other agencies.

'I pray that God will hear my voice and steer it to the hearts and minds of the leaders of the industrialized world,' Bikenibeu Paeniu told the press. 'My expectation is that they may come up with some agreements on reducing greenhouse-gas emissions at this summit.'

I stayed on in Tokyo to try to inject climate arguments into the reporting of the G7's economic deliberations. The attitude of the big seven to Russia was symptomatic of the general state of affairs whenever the finance ministers of the USA, UK, Japan, Germany, France, Italy and Canada got together to talk about economics. The seven ministers had agreed in April to an aid package for Russia which would eventually exceed $40 billion. At the time of the Summit, $610 million of this package had been disbursed. Every cent of it had gone, via the World Bank, to Russia's oil-producing sector. That was what it came down to: a direct link between economic wellbeing and oil production. The people in the finance ministries thought very differently about the world than the people in the environment ministries.

In Tokyo, with the G7 heads of state in town and in every case desperate for good press, I had the best possible opportunity to draw attention to the problems that environmentalists and new economists saw with all this. With the help of a friendly television production company, and posing as a TV reporter, I managed to extract a press pass from the British embassy. With it I could make my way through the security cordons into the holy of holies, the press centre. Here, several thousand journalists milled about in air-conditioned, high-tech splendour, attending one press conference after another. Eager to cast the summit in a favourable light, the ministers and heads of state came to the press in Tokyo, not vice versa.

And the Japanese government had laid on every possible convenience for the journalists while they waited. There were rooms full of photocopiers, fax machines, typewriters, computer terminals and phones. Use of everything was free. Free food was served all day and night in two huge halls, the walls of which were lined with fridge units dispensing free chilled beer and wine. It would be a bold and somewhat ungrateful journalist who wrote anything critical about the Tokyo summit.

Hundreds of NGO representatives had come to Tokyo for the event, many of them environmentalists. Just three of us were inside the press centre. I set about what I knew would have to be a solo task. I typed out an opening press statement and began to run off three thousand free copies on one of the photocopiers. As a security guard walked past, I bent over the tray so he couldn't read the product of his government's largesse to the world's media. It wasn't something the Japanese finance ministry would have approved of. It listed the items on the summit agenda which, in the Greenpeace analysis, threatened economic ruin as a result of climate change. It talked about the threat to the insurance markets, and argued that throwing millions at Russian oil and gas – as opposed to improvements in energy efficiency and new renewable-energy supply – could be likened to lighting a fire, with banknotes, that would in turn burn up billions in global insurance markets. It argued that emerging trade deals, supportive as they were of continued fossil-fuel addiction and environmental deregulation, stood to make a mockery of the commitments world leaders had made on greenhouse-gas emissions at the Earth Summit. It pointed to the opportunities being lost to create jobs by switching to clean-energy technologies.

I worked my way round the press centre with warm armfuls of this press statement, slipping them directly onto the desks of working journalists, or into the hands of talking journalists. It took me two hours to cover every room in the complex. All the releases had my mobile phone number on them in case anyone wanted follow-up comments or interviews.

I went to the eating area, piled myself a plate of free sushi, and sat and waited for the phone to ring.

And waited.

In the next several hours, I had a mere handful of calls. One of the callers was an economics correspondent who wanted to know why an environmental group would bother to come to an economic summit.

Several days of frustration passed. I began to feel like an underperforming used-car salesman. The *Financial Times* man was apologetic, but he couldn't add our perspective to his report. Interest in the environment had gone off the boil because of the recession, he informed me. An agency journalist from UPI, who was one of the few who had phoned and with whom I did a 30-minute interview, told me that his editor had gutted his story. 'They took out all the references to global warming,' he told me. 'The editor told me it is too controversial.' He shook his head. 'You are fighting an uphill battle, you know.'

The G7 economic declaration came out on 9 July. It was written as though environmental issues had no connection with economic wellbeing, or the creation of jobs. It was Bill Clinton's first G7 summit, but there was not a suggestion in the declaration of the handsome environmental rhetoric in evidence during his election campaign.

JULY AND AUGUST 1993, GENEVA

As the G7 leaders deliberated in Tokyo, the worst floods in living memory were building up in the US Midwest. Damage to crops in the nine states affected soon passed the billion-dollar mark. Where a few years before cornfields had languished in severe drought, they now lay below water that covered even farm buildings. Damage to homes and businesses also exceeded a billion dollars. 'I've never seen anything on this scale before,' President Clinton said after touring the area. He asked Congress for $2.5 billion in federal aid.

Towards the end of July, the Mississippi was flooded as far north as Wisconsin. All the bridges along 800 kilometres of river were severed and the total damage bill passed $10 billion. The International Wheat Council slashed its estimates for world grain production in 1993. A grain analyst, speaking to *Time* magazine, explained that the implications of the floods went further than this. 'None of us is ever going to forget how the rains came in the summer for the first time, out of nowhere. And we will never feel the same about our place on earth.'

Calamitous flooding was not confined to America. Over 3,000 people were killed by particularly heavy monsoon rains in Bangladesh, northern India and Nepal. Flooding in China and Taiwan also made the international headlines. On 28 July, one such in the *Daily Telegraph* summarized the situation: 'WORLD WEATHER IN A STATE OF CHAOS'.

As the Midwest floods built up, a paper appeared in *Nature* describing investigations of past climate using cores drilled from the Greenland ice cap. This too made international headlines. It now seemed that when the world was last 2°C or so warmer on global average, more than 100,000 years ago, the climate was prone to catastrophic jumps. Working out ancient air temperatures from oxygen isotopes in the ice, a multinational team of scientists had discovered that the regional temperatures over Greenland had fluctuated by as much as 7°C within as little as a decade. There was only one thing that could change average temperatures that quickly, the scientists had concluded: the switching on and off of ocean currents. Accompanying such switches, it was clear, would have been equally spectacular switches in the patterns of wind, rain and drought. The relatively stable climate of the last 10,000 years, which is to say during the evolution of human civilization, had not been the norm, it seemed, for the warm periods between ice ages.

The implications were obvious, if characteristically understated by the scientists. A reviewer writing in *Nature* came close: 'If the Earth had an operating manual, the chapter on climate might begin with a caveat that the system has been adjusted at the factory for optimum comfort, so don't touch the dials.'

At the eighth session of climate talks in August, governments returned to Geneva for another set of discussions about whether or not to do anything about turning up the dials. President Clinton had announced in April that the USA would be reducing its greenhouse-gas emissions to 1990 levels by the year 2000, in line with the minimum target of all the other industrialized countries. The question now was how the Clinton Administration proposed to achieve that commitment in the short term, in the absence of a carbon tax, and what it would be advocating in terms of targets beyond 2000.

Rafe Pomerance spoke early in the opening plenary. He had

evidently decided to get the bad news over and done with. The Clinton Administration had hoped to bring a new plan with them, he said, but sadly more consultations in Washington were going to be necessary. Pomerance tried to put a brave face on it, but there were fault lines in his voice.

Now that he was head of the US delegation, Pomerance contrived a minimum of contact with his old collaborators in the environmental groups. He was probably being wise. His predecessor, Bob Reinstein, had once explained to me what working for Uncle Sam meant. The only time you could be sure of airing your own arguments on policy was to whatever White House group had been set up on the particular issue, and then only to defend those views against the views of all the other departments with an interest in the matter. On the issue of climate change, that would mean the Treasury, the Commerce Department and the Department of Energy, among others. If your views lost out once the decisions were made, then you followed, advocated and enacted the party line, no matter how far it might be from your own views. You were the White House's servant.

Late one afternoon, with the negotiations over for the day, I went back into the hall to recover some documents I had left on a table. The place was almost empty except for a group of about ten people. Rafe Pomerance was sitting in the USA's position at the tables, surrounded by carbon-club lobbyists. One of these gentlemen had doubtless been responsible for a World Coal Institute briefing which had earlier been distributed to delegates, bearing the title 'Fighting carbon dioxide would lead to economic disaster'. That was probably close to the title of the seminar Rafe was now being given.

Don Pearlman was lecturing the new head of the US delegation. His manner towards Bill Clinton and Al Gore's man in Geneva was as hectoring as I had ever seen him display towards George Bush's. Rafe Pomerance sat there, saying nothing as the fat forefinger wagged at him.

I knew what Rafe believed. I knew he had lost the ability to say what he believed. I wondered if he was happy with his new-found responsibilities. I watched the high priest of the carbon club at his bullying work. What went on in that brain? Did he really believe that there was no threat from an enhanced greenhouse effect? Wasn't he at all worried that what he might be doing was promoting a creeping form of ecological genocide? Did he really believe that doing anything to stop the threat would ruin economies?

He had such power. He, John Schlaes and just a few others were able to orchestrate a multimillion-dollar disinformation campaign with seeming impunity. Journalists seemed to have long since lost interest in the climate talks, but where there was any interest the carbon club was spectacularly successful in throwing up smokescreens. In most countries, market research was showing global warming still well down the list of people's concerns. Much of that had to do, no doubt, with the regular 'global warming debunked' press stories that the Global Climate Coalition and other key carbon-club institutions released internationally, and the regular performances they organized by the handful of scientific sceptics.

Did Pearlman, Schlaes and the rest have consciences? What motivated them? I wished I could understand better.

Most of all, I wished I could find a way to stop them being so effective in their evil work.

SEPTEMBER 1993, WALL STREET

The course of the climate talks may not have been changing, but the same could not be said for the course of nature. Natural-catastrophe insurance losses around the world had reached an all-time record in 1992. Insurers had dubbed 1992 'the year of the cat', yet before 1993 was half over, insurers in the United States had already had to deal with the storm of the century and the flood of the century.

In September, the President of the Reinsurance Association of America, Frank Nutter, elected to become the next major figure in the industry to speak to the press.

He was not a scientist, the reinsurance chief told a press conference at the US College of Insurance, and he did not know whether global warming was to blame for the problems his industry was suffering. But he knew there was genuine reason for concern. The main worry, he said, was this: 'Changes in the number, the frequency and the severity of natural catastrophes are threatening to bankrupt the industry.'

I sat next to Nutter as he spoke, watching the journalists scribbling, adjusting their tape recorders, and fiddling with their cameras, feeling the rare glow produced by a day on which the word 'progress' was offering itself for use in the diary.

Technical questions came and went. I was waiting for the obvious one, and finally it came. Greenpeace is better known for working against industry. Wasn't it strange now to see Greenpeace paying so much attention to the insurance industry?

In fact, I replied, Greenpeace had just as much contact with the oil industry as the insurance industry, maybe more. As we had always told the oil industry, we did not want to put them out of business, we merely wanted them to change the way they did business. We wanted them to become total-energy companies. We wanted them to help make the solar-energy revolution happen, not stand in its way. Our pitch to the insurers was that it was in their own interests to push this case.

The press conference finished, and the television crews set themselves up for individual interviews. Amid all the comings and goings, a video crew from Greenpeace USA was making an eight-minute broadcast-quality film on climate change and insurance for distribution internationally. First before their camera was Eugene LeComte, President of the US National Committee on Property Insurance. LeComte told the Greenpeace camera what his bottom-line fear was. 'Despite the fact that the industry is financially healthy, and has $160 billion in surplus, in two events you could take maybe $70 or $80 billion of that surplus away and you'd cripple the industry. It wouldn't be able to take on new risks. It wouldn't have the capacity to underwrite the business of the future. We'd have massive, massive availability problems.'

Next the video crew filmed Pier Vellinga, a leading Intergovernmental Panel on Climate Change expert on extreme events, and another speaker at the press conference. 'The insurance sector could well be the first victim of climate change,' Vellinga said. 'This might affect the stock market. That is the scenario that many of us fear, that through the insurance sector, the whole financial sector will be affected by climate change.'

I watched my colleagues filming among the journalists. Schemes and dreams flooding into my head, I looked out of the window at the Manhattan cityscape. We were a few blocks from Wall Street.

I did not delude myself that the insurance industry was about to leap on to the global-warming barricades alongside the environmentalists. Among other factors precluding this, right now, was the constant rain of carbon-club disinformation, washing over insurers just as much as the public. A week before the New York insurance conference, the headline in the science section of the *New York Times* had read 'SCIENTISTS CONFRONT RENEWED BACKLASH ON GLOBAL WARMING'. 'Conservatives and industry groups,' a *Times* correspondent wrote, 'have mounted a renewed assault on the idea that global warming is a serious and possibly catastrophic threat. In a drum roll of criticism over the last few months they have characterized the thesis of global warming as a flash in the pan, "hysteria," "scare talk," and a ploy by socialists to justify controls on the economy.' And on the day of the press conference at the College of Insurance, in the letters section of the *Times*, Professor Fred Singer was once again singing for his supper. 'As observations and theory diverge more strongly with each passing year,' Singer's missive concluded, 'it becomes more certain there is something very wrong with the computer models that have been used to scare the world public and their governments into considering drastic, hasty actions.'

NOVEMBER 1993, ZURICH

Swiss Re, at least, didn't seem to share the Singer view. At their Zurich headquarters they now convened a two-day in-house seminar to discuss the ideas in the paper Greenpeace had released after my seminar at Lloyd's of London earlier in the year. As the session unfolded, I realized I was being allowed to witness first-hand a robust debate within the company over what to do about climate change. On one side, a senior manager spoke against any proactive engagement by the industry on the issue. On the other side, Andreas Schraft, the head of the storm and flood group, argued that there was a lot a company like Swiss Re could do. The company would in any event redistribute losses, limit and control capacity, adjust premiums, research, and communicate. It could also optimize clean-energy use in the insurance industry, ask smaller premiums for environmentally friendly technologies, and invest preferentially in cleaner technologies, even if that meant taking a small loss. Why? First, because in the long run insurers would make less profit anyway if climate change hits. Second, because, as Schraft put it, 'the insurance industry has a vital interest in the survival of mankind. It won't help us if the world disappears that we have made a profit.'

The discussion ebbed and flowed around the two opposing viewpoints.

After the discussions, I was taken to see a member of the Swiss Re board. Walter Kielholz was a dapper man with an office the size of a Zurich bank vault. We chatted pleasantly for fifteen minutes. Kielholz, who would later become chief executive officer of Swiss Re, scrutinized me from behind his mammoth desk without giving anything away. He had earlier told Swiss television that his industry should do something about climate change, but he certainly wasn't going to elaborate in front of me. I had a sense, in talking to him, of the problems facing a board split between the opinions I had heard his colleagues espouse. The net result of the stalemate, of course, would be a continuing absence of proactive action.

JANUARY 1994, SUSSEX

In November 1993, with Greece suffering its worst drought for a hundred years, Greek religious leaders asked the population to pray for rain. Many of them were otherwise occupied trying to breathe. The prolonged heat had compounded another dire environmental problem – air quality in Athens. As air pollution reached record levels, and with over a thousand in hospital with consequent respiratory and heart problems, the government banned a million cars from the city centre.

In California, meanwhile, two weeks of brush fires were in the process of clocking up another 'billion-dollar cat' for the insurance industry. Twenty-six firestorms burned some 800 square kilometres from the Mexican border to Los Angeles.

In December the Greek investment in prayers was repaid with compound interest. The resulting floods blocked highways and swept away cars. It was a rainy month in most of Europe, and over Christmas the worst floods in sixty to a hundred years spread across northern France, Belgium, the Netherlands and Germany. With scores of riverside towns and cities affected, thousands of Europeans were evacuated, many spending their Christmas in army barracks.

I spent my Christmas on dry land, eating mother's home cooking and reading Paul Kennedy's *Preparing for the Twenty-first Century*. The famous Yale historian had turned futurologist. He had an entire chapter on the environment, and it was dominated by gloomy thoughts on global warming. This was one of the great dismaying trends of the twentieth century, which Kennedy used to paint his picture of the twenty-first. 'The forces for change facing the world could be so far-reaching, complex, and interactive that they call for nothing less than the re-education of humankind,' he concluded. This was precisely what many social thinkers had argued before, a race between education and catastrophe. But now with a clock ticking too loudly to ignore.

A fool, I read somewhere once, learns from experience. A wise person learns from history.

The year 1993 proved to be the sixth hottest on record. The hottest nine years in 140 years of record-keeping had all now been since 1980. With its $62.5 billion of economic losses from natural disasters, 1992 had been the worst loss-year ever; 1993, with $50 billion in economic losses, $10 billion of them insured, had been little better. Neither did 1994 open promisingly. The first week of the new year saw over a hundred and fifty bush fires raging in New South Wales, and Sydney itself under threat. Around 20,000 people were evacuated from suburbs while 7,500 firefighters, backed by the military, battled to keep the flames out of the city.

How much longer would it be before the wider world would wake up to the wider flames at the wider city gates?

FEBRUARY AND MARCH 1994, OXFORD

The Intergovernmental Panel on Climate Change was by now examining the question of the potential impacts of global warming on the financial sector. The IPCC's First Assessment Report of 1990 had given the impression that there would be no impacts on insurers, banks or pension funds. But by the end of 1993, a group of insurance and banking experts had been set up to write a chapter of the IPCC Second Assessment, due for completion in 1995, on the impacts of climate change on the financial institutions. A top British insurer, Dr Andrew Dlugolecki, had been appointed as lead author. Dlugolecki, manager of operations with General Accident, had a long history of interest in climate change. As long ago as 1989, he had served on a panel appointed by the British Government to assess the impact of climate change on the UK.

In February 1994 Oxford University held a seminar on the effect climate change might have on the British economy, aiming to attract high-level participation from the British Government. The venue was Green College, an institution renowned for its environmental policy work largely by dint of having as its head Sir Crispin Tickell, environmental advisor to both John Major and Margaret Thatcher. For the occasion, Andrew Dlugolecki and I were invited to present a joint lecture on the impact of climate change on the financial sector.

We had very little time to liaise ahead of the event. Dlugolecki proposed over coffee that, by way of a change, I should present only the data, and he should offer the interpretation and any rhetorical conclusions that might be necessary.

It must have been strange indeed for the assembled Treasury officials and others to see a Greenpeace campaigner giving a dry account of windspeed/damage ratios and the implications of coupled climate models for storminess, all devoid of rhetoric and with the aid of multicoloured overhead projector slides complete with a corporate logo – General Accident's, not Greenpeace's. Then Dlugolecki gave his overview. He began by describing how, as a member of the British Government's Climate Impacts Committee, he had spent a whole day in the Department of the Environment in 1990 finalizing the committee's report, only to come out and see a city brought to a virtual standstill by the second most extreme weather event in recorded British history – one of the four billion-dollar European windstorms that had so worried the industry in January and February of that year. This coincidence, however, was not the only reason why he felt that climate change was a potentially huge threat to his industry.

'I find it quite amazing,' he concluded, 'that the insurance industry has no body to represent its interests on climate change to the international community.'

As for the other sectors in the financial institutions, he had spent some time already trying to interest the banks in the issue. In a sense, he said, the insurance industry provided cover only for a year at a time. The banks offered loans over periods of up to twenty years. Many of these loans were for projects themselves vulnerable to the impacts of climate change; others were for projects that would only boost the greenhouse effect, and augment the risk to the financial sector. Yet he had found the banks to be completely uninterested in the issue.

During the first three months of 1994, further intense storms battered the east coast of the USA. Even *Time* magazine now started to think about climate change. 'BURNED BY WARMING' announced the headline of a full-page article. The journalist who wrote it had been at the College of Insurance press conference, but had not been able to persuade his editors to run a story on the issue at the time. He now described how the US insurance industry faced a two-pronged threat. First, and increasingly clearly, there was the suspicious intensity of storms such as those in the nastiest winter in living memory. But second, storms or no storms, the sea-level rise predicted in the years ahead would threaten $2,000 billion worth of insured property along the Atlantic and Gulf coasts. 'These risks,' the article concluded, 'and the crucial role played by the $1,410 billion insurance industry in the global economy could change the dynamics of the debate about global warming.'

The same week I saw those encouraging words, I learned of perhaps the most depressing scientific discovery yet. Dr Peter Wadhams of the

Scott Polar Research Institute told a major oceanography conference that convection in the Greenland Sea had virtually stopped in the last decade. This he had discovered as a result of the latest of several cruises he had led over the last ten years to investigate a prominent tongue on the fringe of the Arctic ice cap, where ice formation causes dense salty water to form, which in turn sinks and drives ocean circulation. When ice forms, salt is excluded, making the residual seawater saltier than normal. This ice-related, density-driven circulation operates only in three small areas of ocean: in the Greenland and Labrador Seas in the Arctic, and the Weddell Sea in the Antarctic. A fourth area of density-driven circulation, related to evaporation rather than ice, operates where the Mediterranean joins the Atlantic.

On a cruise ten years before, Wadhams had measured convection in the Greenland Sea descending to the seafloor 3,500 metres below. Five years later, it was down to under 2,000 metres. Last year, it had been barely 1,000 metres.

'It is a process that has been going on for several thousand years, and when you see it decline sharply over a decade it gets you worried,' Wadhams told the UK press, with a scientist's typical understatement. That less ice was forming in the region was something he thought was probably a function of global warming. What was more, the reduced convection would reduce the amount of carbon dioxide that could be taken down into deep water – normally about 25 per cent of the global total sequestered in this way, Wadhams estimated – and so would boost global warming still further.

Reading of this in the UK national press, I thought of Dan Reifsnyder sitting in the State Department in Washington. This scenario, I recalled, was one which he had told me would – if it had transpired – awaken Washington, and therefore the world, to the threat of global warming.

I had doubted it then, and I duly watched the few column-inches in the newspapers – having failed to ignite a burst of high-level meetings in the White House – turn into recycling fodder the next day.

6

Excess of Loss

March–December 1994

MARCH 1994, ANCHORAGE, ALASKA

On 21 March 1994, the Framework Convention on Climate Change came into force. The terms of the treaty required 50 nations to ratify the convention before it became an instrument of international law. The 50 ratifications had been clocked up many months earlier than the precedents of international treaty-making had led everyone to expect.

On a plane flying from Seattle to Anchorage, I bought a bottle of champagne and allowed myself a quiet celebration. The convention becoming law was not something I was going to be able to read about in the US newspapers. And there were no teeth in the convention, no legally binding clauses referring to specific targets and timetables for action. But at least we were on a road where the danger ahead had been recognized. It was the beginning of a process.

For the present, there was another environmental anniversary which I knew I *would* see covered in the press. In a few days' time it would be the fifth anniversary of the *Exxon Valdez* oilspill.

I looked out of the right side of the plane across the sea at the Chugach Mountains, rocky teeth and snowy drapes against a blue sky. I followed the coast on my map, checking landmarks as the plane flew towards the area where Exxon's tanker had foundered. I spotted the fishing town of Cordova, where many of the Alaskans in litigation with Exxon lived. Then the plane was over Prince William Sound.

On the north shore of the sound, the Columbia Glacier tumbled into the grey water, all crinkles and crevasses. Near its serrated front, cusps of white drifted. It had been the threat of icebergs from the glacier that had forced the *Valdez* to alter its normal course on the night of the grounding and move closer to the rocks. Icebergs had this very week caused the port of Valdez to be closed to tanker traffic.

As I neared Anchorage, I wondered what the next week of my life was going to be like. My colleagues and I had the brief of trying to maximize Exxon's discomfort on the anniversary, and seeking to make a link between the company's behaviour over the oilspill and its behaviour over climate change. The first bit, at least, looked as though it should be easy. The reason for the accident seemed clear. As the Alaska oilspill coordinator had put it, 'there were simply arrogant and complacent people at the top levels of Exxon Shipping Company and Alyeska Pipeline Service Company, who did not pay attention to their responsibilities.' Exxon stood guilty of criminal acts as a consequence. It had pleaded guilty in October 1991 to criminal charges arising from violations of the Federal Water Pollution Control Act and other Acts. The company had been fined $150 million. It had also agreed with the state and federal governments to pay a fine of $900 million over ten years to cover civil offences.

But in Anchorage I had to try to argue that what Exxon was up to in terms of peddling disinformation over climate change was also complacent and irresponsible, if not criminal. I rehearsed the chronology in my mind for the press conference the next day. Immediately after the spill, in April 1989, Exxon's then chairman, Lawrence Rawl, had offered his assurance that within a year the toxic pollution would be eliminated. Two years on, in April 1991, the US National Oceanographic and Aeronautical Administration (NOAA) had reported that the environmental damage would persist for much longer than scientists originally thought, and in some cases would be permanent. None the less, in Exxon's 1991 Progress Report on the Environment, Rawl and the then president, Lee Raymond, had insisted that there was 'strong' evidence that Prince William Sound had 'essentially recovered'. But in April 1992, the Oil Spill Trustees – the body set up by the federal and state governments to allocate the fines imposed on Exxon – issued a long report on the impact of the spill which concluded, among other things, that commercial harvests of pink salmon had declined, and that organisms continued to be exposed to carcinogenic hydrocarbons from subsurface oil on beaches. In the face of this, Exxon's 1992 Annual Report concluded that recent surveys confirmed that 'Exxon's massive clean-up effort, together with the natural recovery process, has left shorelines essentially clean and recovered.' However, in February 1993, NOAA, the US Fish and Wildlife Service, and the Alaska Department of Fish and Game released the results of four years of monitoring at a symposium in Anchorage. They showed a depressing catalogue of environmental damage since the spill. The toll included harlequin ducks failing to breed in any year since the spill; brain damage in seals similar to that generated by solvent abuse in humans; low otter populations in 1989, 1990 and 1991, and unusually high mortality among otters in their prime; late and disrupted breeding among guillemots; genetic damage in herring; and reservoirs of oil still trapped in gravel and mussel beds.

Exxon should by this stage have been finding it increasingly difficult to stick to their black-is-white, the-earth-is-flat line. But in April 1993 the company finally presented the results of its own monitoring of Prince William Sound, at a symposium held far from Anchorage, in Atlanta, Georgia. The contrast between the Exxon version of what was going on in Prince William Sound, and that of the federal and state government agencies, could not have been greater. Alan Maki, Exxon's chief scientist, told *Scientific American* that the company felt the sound had 'essentially recovered'.

Two Cordova residents, both scientists as well as fisherfolk, travelled to Atlanta to bear witness that they and a number of others thought otherwise. Drs Riki Ott and Rick Steiner had become famous by 1993 as a result of their role as Exxon watchdogs. 'Exxon is constructing a false reality of the situation in Alaska in order to influence public perception,' they wrote in a paper prepared for distribution at the Atlanta meeting. 'They are intentionally and flagrantly understating the extent of the damage for obvious reasons. Because Exxon has tens of billions of dollars at stake in outstanding civil lawsuits over the spill, it has a huge vested interest in how both the spill and recovery from the spill are portrayed. By manipulating data and selectively presenting information, Exxon has constructed its own reality of the spill.'

Given the facts, I should have been expecting an easy ride in Alaska. But my colleagues at Greenpeace's Alaska office had warned me that this was oil territory, and that I would be fooling myself if I thought Exxon was necessarily against the ropes here. The company had a lot of supporters in the state government, in the press and elsewhere.

The next day, the turn-out at the press conference in Anchorage was disappointing. Two of the three state TV channels were there, but no CNN or US network crew, and no agency journalists. Before we even started, it was clear that we would fail to carry our message beyond the state. The most important paper in the state, the *Anchorage Daily News*, had not even turned up.

Riki Ott proved to be a sparky and competent woman, a Ph.D. in toxicology who, seeking the Alaskan wilderness life, had ended up as a Cordova fisherwoman. She had prepared a paper for Greenpeace summarizing the latest findings on the after-effects of the spill. She gave a fluent summary of the growing casebook of depressing evidence, which now extended to an emerging link between genetic damage and reproductive impairment in both herring and salmon.

I then had to try to make the leap from the oilspill in the sound to the 'oilspill in the sky', as our rhetoric put it – the link between oil use and climate change, and the way Exxon were lying about both the effects of the oilspill and of oil use.

The press conference finished without any feedback on how successful I had been. The local TV crews positioned their cameras, and journalists, to film interviews. A very young journalist pointed his microphone at me. His tone verged on the sneering. 'Can you give me some examples of the climate changes this oilspill caused?'

I took a deep breath. I tried again to explain, this time on camera. A frown replaced the sneer. Give me sound bites, not a seminar, I could see him thinking.

A journalist from a second channel wanted to interview me outside. Like a fool, I agreed. He took an age setting up, and then asked me about the impact of global warming on the Arctic. This question I had to answer through blue lips, my breath visible as frosty clouds against a backdrop of piled snow, visibly fighting shivers.

The following month, Al Gore gave a major speech on climate change. It was Earth Day, and the US Government had recently – finally –

announced its action plan on climate change. The package was pathetic. There were no measures to deal with the rapid growth of emissions projected in the transport sector, no real stimuli for energy demand-side management, no efforts to redress the huge public subsidies given to fossil-fuel exploitation. Gore none the less had to try to dress the plan up as progress.

He delivered his speech on the campus of Georgetown University in Washington. The vice-president reaffirmed his view that climate change was the number one threat to the human future. Moreover, he said, it could have ruinous impacts on economies in the short term. Take a recent quote, he told his audience of policymakers, environmentalists and students. 'The insurance industry is first in line to be affected by climate change. It could bankrupt the industry.' Where had he got this, Gore asked. From the President of the Reinsurance Association of America.

APRIL 1994, LAGUNA NIGUEL, CALIFORNIA

The 1994 annual meeting of the Reinsurance Association of America was into its first morning. The professor from Pennsylvania opened with a joke about baldness. He himself was bald, and so were many of the 28 chief executives listening to his presentation. I wondered if he was being sensible. I sat at the side of the conference suite, so I could watch how the chief executives reacted to the professor. It was my turn to speak to them next, and the professor and I had very different stories to tell.

The executives assembled that April day in Laguna Niguel's Ritz Carlton hotel represented more or less the entire American reinsurance industry. The world's insurers had paid reinsurers some $100 billion for cover against catastrophes damaging property the previous year, much of it to the companies represented in this room. Little wonder that America's reinsurance bosses had chosen to review their catastrophe business in one of California's most expensive pieces of real estate.

Outside the hotel, surfers rode waves on one of the Sunshine State's most famous beaches. Inside, the chief executives were about to consider whether or not global warming would mean a world in which it was harder to make money.

In making my case, I was not going to be helped by the fact that the rubber-suited figures in the surf outside were shivering in some of the coldest spring weather in years. Neither was I going to be helped by the professor from Pennsylvania. The chief executives were supposed to listen to a half-hour counter-view from him, then half an hour from me. They would then have fifteen minutes in which to grill us both before heading off to another of their gourmet lunches. The more convincing the sceptical professor was, the harder my job would be. I couldn't afford to spend too much of my thirty minutes persuading the executives that global warming was a major threat to their business. I very much needed to move swiftly to my main case: that there were ways for the insurance industry to organize itself to help prevent the threat in the first place. Or at least head off the worst of it.

The executives laughed at the professor's baldness joke.

I listened to the professor build his case, his manner easy and credible. He came to the first of the key arguments of the global-warming sceptics. Variations in the strength of incoming radiation from the sun were more important in determining global average temperature, he explained, than the build-up of greenhouse gases in the atmosphere as a result of the burning of oil, gas and coal. I had never heard of the man before, never seen a paper on climate change written by him – nor one even referred to in other papers. He was a regular consultant to reinsurance companies on clean-up liability, I had been told.

I looked at my watch. He would have to speed up. After an over-long introduction, and two lengthy jokes, he was well into his thirty minutes. He had barely started the litany of arguments I had come to know from his kind of scientist.

I swallowed my frustrations. I tried not to think of the times I had seen these arguments rebuffed by the scientists from NASA, the Meteorological Office, and all the other government and academic institutions represented on the IPCC. The professor from Pennsylvania had not been part of that unprecedented scientific advisory process. He believed he knew better than the three-hundred-plus scientists who had thrashed out the IPCC's scientific reports of 1990 and 1992, and compiled a scary report on radiative forcing soon to be published in 1994 complete with the latest calculations of how much heat was trapped by each type of greenhouse gas. In fact, as I was about to find out, he was not even going to mention the Intergovernmental Panel on Climate Change.

I looked at my watch again. The professor had not once looked at his. He was relaxing into his task, weaving other little jokes into his critique. He was enjoying himself. He had only five minutes more, and I knew he had a lot left to say.

I looked around the room for Frank Nutter. His duty was to marshal the chief executives, and their various spouses, through their two-day stay at the Ritz Carlton. He was bent over an administrative task at the back of the room.

I could still barely believe that Nutter had invited me to this occasion. I was here not only to present my case to his members, but to stay with them in their luxurious base for the full duration of their conference. I would even be playing golf with them the next day. It was a lobbyist's dream – months of normal corporate lobbying opportunities condensed into two days, and among those right at the very top of a key industry. As days went, this was a big one for me.

I shuffled through the paper I had prepared to accompany my presentation, and tried to focus on the potential pot of gold I was aiming at. The Chartered Insurance Institute in the UK was just about to release a report on global warming which summed up the state of play perfectly. 'The industry has a limited breathing space in which to gather its wits, and plan in a truly long-term timeframe,' it concluded. And among the recommendations would be the first ever advice from within the financial sector that investments were part of the problem, and hence could be part of the solution. The final sentence of the report gave the bottom line: 'All investment managers should modify their investment policies to take account of the potential direct and indirect effects of global warming.'

The day that happened, I figured, could well mark the beginning of the end for oil and coal. Investments by insurance and reinsurance companies were measured in hundreds of billions each year. These companies, along with banks and pension funds, owned over half the equity in corporate America. Start switching that crowd on to solar, and away from carbon, and it would be the beginning of the end for the threat of a destabilized global climate, the end of the nightmare of a planet slowly creeping towards a state too hot to live in. Into the bargain, it would also be a major step towards heading off other environmental problems, such as acid rain, urban air pollution, oil-spills, and the incremental leakage of toxic pollution from refineries and the rest of the oil industry's paraphernalia.

A pot of gold indeed. But the professor from Pennsylvania was now lampooning Al Gore.

'Gore,' the professor was saying, 'invokes in support of his theory exactly the piece of evidence that most directly refutes it.'

The chief executives laughed.

'I haven't had the chance to discuss it with him.'

More laughs.

I realized that my teeth were so firmly clenched that my jaw was beginning to ache.

I thought about our location. I been driven here past blackened, burnt-out hillsides. Huge bush fires in November 1993, just four months before, had spread westward to engulf the outer suburbs of Laguna Niguel. The driver had noticed my avid scrutiny of the terrain. He described the night the fires erupted out of the canyon. He and hundreds of other citizens had joined the fire service to fight them until dawn.

The six-year drought in California had had much to do with the ferocity of the fires that night. Laguna Niguel had been one of the areas worst affected by what in fact turned into the fourteenth weather-related 'billion-dollar cat' for insurers since 1987. The irony of the location of the meeting was not lost on me, and I hoped it would not be lost on my audience either – when I eventually got the chance to speak to them.

The professor from Pennsylvania was now well over his half-hour limit, and cutting into my time. I looked around for any sign that someone would remind him. But everyone seemed engrossed. He was now attacking the computer climate models on which the IPCC based their estimates of future rates of global warming.

Normally, as the time approached to give a speech or presentation,

I would feel my nervousness slowly give way to a state of calmness. It was not happening today.

The professor from Pennsylvania rumbled on, as relaxed as if he were with a class of undergraduates. He started talking about energy policy, and showed a string of slides illustrating upward growth in oil, gas and coal consumption. Nobody showed any sign of waning patience.

My glances at the clock became more and more frequent.

I could see the minute hand moving.

The professor finally built to his climax. 'Avoid,' he advised the chief executives, 'the trap of using the fear of global warming as a whip to change behaviour.'

They clapped.

He sat down.

I stared at my watch in disbelief as Frank Nutter ran quickly through my history, and the reasons why he had invited me to speak to the Reinsurance Association of America. The professor from Pennsylvania had taken 63 minutes of the 75 allocated to the session on the programme.

I walked to the podium with a dry mouth, clutching my carefully prepared overhead projector slides in fingers wet with sweat. I looked at the audience. A wave of pre-lunch fidgeting greeted me. Michael Butt, the suave chief executive of Mid Ocean Re, whom I had visited to lobby in his stronghold on Bermuda, was reading something seemingly unconnected with the events at hand. Stephen Gluckstern, the President of Centre Re and the widely acknowledged financial *wunderkind* of the industry, was staring at the crystal chandelier, lost in some inner thought. The representative of Munich Re, the largest reinsurance company in the world, was frowning.

I thanked Frank Nutter for his introduction, and tried to shrug the invisible weights off my shoulders.

'It is true that I have a rather different view to offer you,' I began, echoing Nutter's introduction. 'And I would love to have 63 minutes in which to do it, while at the same time rebutting many of the points made in the previous talk. However, according to the programme, I now have only six minutes. Assuming, that is, that we abandon the time for discussion.'

This was greeted with a half-hearted laugh from the executives, who seemed to be looking everywhere but at me. Frank Nutter called out from the back of the room that I should take my time. But I could hear dinner plates clattering faintly in the distance.

The situation reminded me of a dream from my childhood. Everyone is anxious to head off somewhere important, but I am seemingly rooted to the spot. I have something important to tell the imminent departees, but I can't seem to make them listen. Neither can I quite articulate what it is I want to say.

I realized that my thoughts were now trailing several milliseconds behind my words. That hadn't happened in fifteen years, since I had given some of my earliest and worst university lectures. My mouth was drying up, as though all the moisture in it was required in my palms. I realized I was going to have to stop, and use ten of my precious seconds to take a drink of water.

Guys, listen, I want to talk to you about a threat to life on Earth. I want to tell you how you can help stop it. I want us to consider together how we can stop making the capital markets an engine of planetary suicide, and turn them into an engine for planetary survival. We have to bring about the fall of the carbon era, and the rise of the solar era. Nothing else can do it.

But all I could do was think it. 'Er, in the next overhead, you can see . . .'

It was a nightmare.

I stumbled on, sleepwalking my way through it.

I had the rest of the two-day gathering to try to assess the damage, and do some repair work. Unlike the professor from Pennsylvania, who had lobbed his careless grenades and departed, I went to every available cocktail party, dinner and late-night session in the penthouse hospitality suite. My liver and cheek muscles became key servants to the cause. And the going was tough. The executives were there with their spouses. I was clearly in grave danger of being even more of a greenhouse bore than was normally the case.

I did what I could, seizing each and every opportunity to supplement the case I had partially made in the formal session, meanwhile trying to be sociable as the conversations swirled through the state of the bond markets, or who the Republicans should put up against Clinton. On the golf course, I risked the occasional greenhouse comment as I waited with three executives on the tees, before invariably driving allegorically into rattlesnake-infested natural terrain, rather than the herbicide-drenched manicures otherwise known as fairways.

The US arms of Munich Re and Swiss Re were represented at the meeting, and here, at least, the ground was fertile. Their chief executives knew the direction in which things were going at their European headquarters.

The Americans were more of a problem. 'Who do we believe?' one executive said to me, shrugging. 'Just like at this meeting: you get some scientists saying one thing, and others saying the other.'

I nodded, looking up at the fire-blackened hillsides where the southern California wildfires of 1993 had lapped against the millionaire suburbs where we stood.

JUNE 1994, CITY OF LONDON

On 14 June I opened my *Financial Times* and saw that the court case in Anchorage had produced a first result after just six weeks: much quicker than anyone had expected. 'EXXON GUILTY OF OIL SPILL RECKLESSNESS,' the headline read. A second phase of the trial, with the same jury, would consider the scale of punitive damages. The fisherfolk and others involved in the class action sought $15 billion.

Exxon's share price lost 4 per cent immediately. Three billion dollars were wiped off its market value, and within a week the two top US rating agencies, Standard & Poor and Moody's, had downgraded Exxon's credit rating. The company made $5 billion profit in a good year; now they faced the potential loss of three years' profit in one class action, for one act of pollution. In fact, when the Alaskan court finally announced the damages Exxon would have to pay, the sum would be $5 billion, not $15 billion. But the lesson was still there for all to see. *The Economist* spelt out just how big the liability stakes had become for big oil, with the emergence of the Oil Pollution Act and other liability developments. 'Enormous as the sum might seem, it is nothing compared to the damages that a polluter would face under the act that Congress introduced after that oilspill.'

A few days before Exxon's indictment, two pillars of the British establishment met at a function. One was the editor of the *Sunday Telegraph*, the other the chief executive of ICI. The newspaperman was amazed to hear in passing from the boss of the chemicals giant that the top item on the agenda when ICI was next to hold its quarterly meeting of divisional bosses would be the environment. This much impressed the man at the helm of the UK's most conservative Sunday newspaper. When, within days, Exxon hit the rocks of a potentially multibillion-dollar recklessness charge, he was even more impressed. It seemed that the environment now had a passing chance of going top of the agenda in oil companies as well as chemical companies. His response was to ask his city editor to write a major feature on liability.

The latter did his research, and wrote his article. He talked at length about the payouts required of oil companies as a result of the US Superfund legislation for toxic liability, and the strong prospect that such a regime would spread to Europe. He even raised the spectre of punitive liability payouts tomorrow for profligate emissions of greenhouse-gas emissions today. I particularly enjoyed his closing quote, which came from an environmental lawyer: 'European companies must understand, in the way their US counterparts have already learned to their cost, that full-scale attacks orchestrated by environmental groups are positively to be expected on whole industries. Decisions to invest in industries which are then subject to such attacks will also be heavily criticised and we may indeed see test cases brought by pension holders against fund managers on environmental grounds. Today's fringe group fantasy may become tomorrow's fund managers' nightmare.'

AUGUST 1994, BRISBANE / SYDNEY

Running through Brisbane's botanical gardens at 7 o'clock on a clear-skied morning, it seemed difficult to believe that anything could be amiss with the world. Representatives of what seemed like every species of palm on the planet cast long early-morning shadows across the dew-glossed lawns. A towering banyan tree stood starkly green against a silver backdrop of glass skyscrapers in the city's commercial district. Fresh air filled my lungs, and for the first time in years I could feel real strength in my legs.

Two months earlier, my close colleague and head of delegation at the climate negotiations, Bill Hare, had suffered a stroke. He was thirty-eight. It was ageing, chain-smoking, overweight fossil-fuel lobbyists who were supposed to have strokes, not young, non-smoking, slim environmentalists, albeit workaholic ones. Bill, an Australian, was widely regarded as the best climate lobbyist in the world. He had a unique quality, in that he was on the one hand a trained atmospheric scientist able to engage climate scientists in debate at the highest level, and on the other – rare for a scientist – possessed of a prodigious political acumen. He was able to run rings around most economists, and had done so many times on live TV in Australia. He was supposed to have been here in Brisbane with the Greenpeace team, not struggling to come to terms with what had been a near-fatal assault on his health. We were at the 1994 South Pacific Forum meeting, and his country – the host country this year – seemed to be in the process of caving in to its fossil-fuel lobby, and reneging on its commitments under the climate convention. We were missing Bill in more ways than one.

The sky may have been blue, the gardens fragrant, and the city glittering, but all was not well with the world hereabouts. Not far from these botanical gardens, Queensland farmers were enduring the worst drought in living memory. I had come here from New Zealand, where the north of the country was also suffering its worst drought of the century. Signs by the roads in Auckland had read 'No water. No pubs. Save our city. Save water.'

Meanwhile, out in the Pacific, Polynesia was enduring its worst ever episode of coral bleaching. The deadly white rashes on the reefs extended from Tahiti to Samoa and the Cook Islands. Polynesia had already suffered very bad bleaching episodes in 1984 and 1991. They had barely begun to recover from the 1991 event.

I stopped, panting, at the crest of a rise. A primitive tree stood before me. Ten metres high, it had simple branches radiating from a central node, bifurcating at regular intervals, and capped by a flat fringe of oblong leaves the size of rugby balls. It looked ridiculous, more like a child's drawing of a tree than a tree itself. I walked up to it. A label told me it was a dragon tree, and that it could live for three to four hundred years. I read on. The tree had been transported to Australia in 1862 from the Canary Islands, and was one of only two in the whole of the country.

This ridiculous little tree had been alive three years before John D. Rockefeller had formed the world's first oil company.

The tree, I read, had lost a third of its branches in a storm in 1984. 'Please look after and respect our unique dragon tree,' the label entreated. I ran on again, trying to control the anger I suddenly felt.

Down the gardens, close by the hotel where the heads of state were staying for the South Pacific Forum, a huge hot-air balloon hung in the still morning air, ropes anchoring it in place thirty metres off the ground. Its canopy was painted as a globe. A message sprawled across the blue seas and green continents: 'Stop Global Warming'.

Australia had been preparing for its retreat for a long time. The pressure from the carbon club had been building steadily ever since the Cabinet decision, back in 1990, to set a target of 20 per cent cuts in carbon dioxide emissions by 2005. The Business Council of Australia, dominated by mining and oil companies, including Shell, had led a visceral and sustained political backlash. Global-warming sceptics, including Richard Lindzen and Fred Singer, had visited Australia to do the rounds of radio, TV and print media. The regularity of 'global warming debunked' stories, and the degree of vitriol in the debate, were such that it seemed to my Greenpeace Australia colleagues as though their country had somehow been singled out by the carbon club as the weak link among the nations with emissions targets.

That would have made sense, of course. Australia, the world's leading coal exporter, was easily the most resource-dependent economy among the countries with emissions targets. I had done tours for my Australian colleagues every year since 1990. In Canberra, despite the government's policy, the only minister I ever met who seemed to support action on climate change was the environment minister. In Melbourne during 1992, at the Business Council of Australia's sumptuous headquarters, I had faced the most implacable and hostile group of industry representatives I had ever encountered. These people were much more frequent visitors to Government House than environmentalists, and they visited with somewhat more weight that they could throw around.

In 1994, with talk in the air of the need to strengthen commitments at the climate negotiations, the Business Council of Australia launched an all-out assault on their government's target. The cost of achieving 20 per cent cuts would be as much as 40 billion Australian dollars, they announced. The target should be abandoned. If ever coal-dependent Australia signed up to another target on greenhouse-gas emissions, it should be set at a level lower than in the other industrialized countries.

A week later, the foreign minister warned that Australia might refuse to discuss greenhouse-gas reduction commitments at all. The Cabinet had actually discussed the option of not accepting commitments on climate change, he admitted. The government had only ever put forward its target as an interim planning measure.

A disastrous failure of nerve seemed to be brewing. Australia was clearly in danger of being the first country to break ranks with the spirit, and perhaps even the letter, of the Convention on Climate Change and its commitments. To do that at any time would have been calamitous. To do it just as Pacific island nations were drafting a protocol to the convention aimed at tightening commitments, and doing so with reasonable hopes of support from countries like Denmark and Germany, would be treacherous. To do it just as drought and other potential early signs of a climate in crisis were breaking out all around the world could even be suicidal.

We launched our campaign on the first day of the forum. Pene Lefale, my Samoan colleague, laid out our basic case to a packed press conference. I watched Pene speak, as I had so many times before, in so many places, in the previous four years. His voice, softly belying his rugby-player's frame, still gave away the tiniest undercurrent of the emotion that the threat to his homeland raised in him. He explained carefully the promises Australia had made, the disastrous chain of events that Australia would set off at the climate talks – soon to reconvene in Geneva – if they backed off their target. And this was not an abstract argument about some uncertain future threat, he insisted. The first probable victims of climate change were already entering the hospital wards.

Next to Pene sat Dr Ove Hoegh-Guldberg, a University of Sydney coral biologist. He had spent the last month on contract with us in Tahiti, carrying out a survey of the stricken reefs, accompanied by a film crew. Hoegh-Guldberg began his presentation with a video. It was only eight minutes long, but it had made me feel physically sick when I had first seen it.

I watched the sombre faces of the journalists as the underwater images flickered at them. Hoegh-Guldberg had found bleaching at every site he had dived on. Even at 25 metres the coral was still bleached. He expected that at least half of the coral currently alive around Tahiti would die before the episode was over.

The video moved on from the whitening staghorns and hovering divers to pictures of satellite maps showing water temperature. There was no doubting it, Hoegh-Guldberg said: the water was several degrees warmer than it should have been for the time of year. That was why the coral was dying.

The Pacific island leaders were in no mood to tolerate a betrayal by Australia in Brisbane. At the Earth Summit in 1992, the small island states had been promised a future summit all of their own. The UN Conference on Small Island Developing States had taken place in Barbados in May. The prospect of this event had done much at the Earth Summit to hold back the despair that had been voiced by the island nations at the World Climate Conference in 1990. They had a forum to look forward to at which they could hope that their concerns about global warming would be addressed. But as I had seen at first hand as part of the Greenpeace team on Barbados, the Small Islands Summit had delivered nothing. For Australia now to dismantle the usual alarming references to global warming in a South Pacific Forum Communiqué would be difficult indeed.

The communiqué issued by the governments at the end of the forum duly reminded its readers that global warming and rising sea levels were 'among the most serious threats to the Pacific region and the survival of some island states'. The next sentence was as good as the diplomats could manage in trying to match the growing panic in the Pacific with emerging government reluctance in Canberra even to freeze emissions. 'The Forum recognised that existing commitments in the Framework Convention on Climate Change will not meet the ultimate objective of the Convention and called for early agreement on a process for negotiating one or more protocols to implement and elaborate the Convention, so that reductions of greenhouse-gas emissions are achieved.'

Two weeks later, Australia went to the next session of climate talks in Geneva with its emissions target still in place. Sources within the Australian delegation declared themselves secretly delighted with the pressure their government had come under.

We had won another small rearguard action in the carbon war.

I left Australia, next stop Japan. In the transit lounge at Cairns Airport, I watched the national television news. The lead item came live from a coal mine. The screen filled with images of mine workings, and stricken faces. The Moura mine in central Queensland had suffered its third disaster in twenty years. An explosion had trapped eleven men over two hundred metres below ground, and three kilometres from the entrance of the mine.

At this same mine, thirteen men had died in an accident in 1975, and twelve in 1986.

I bought *The Australian*, and read the details of the tragedy. As a geologist I had only ever once been down a coal mine. I had lain at the coal face, a kilometre and half below South Wales, shaking with fear as flakes of the roof rained down on me while grinning Welshmen adjusted pit props.

'Underground mining,' a United Mineworkers spokesperson had told *The Australian*, 'is a dangerous occupation.'

AUGUST 1994, TOKYO

I had visited Tokyo in August many times, having worked with research collaborators at the University of Tokyo and elsewhere in my previous life, so I knew I was experiencing a singular heat the moment I got off the plane. The heatwave, I quickly learned, had been protracted. Tokyo's main water supply, the Tone reservoirs, would dry up by the end of August if the heatwave continued. Incredibly, Japan had been forced to import water.

First Auckland, then Queensland, and now Japan. Each of my ports of call on this trip was in the grip of its worst drought of the century. It was becoming farcical. Things were so bad that the Japanese Government had abolished controls on the import of bottled water, and was in the process of relaxing health standards to widen the brand names that could be imported. Oil refineries had also imported water. Japan Energy, a major refiner, had had to import 31 million litres of water from South Korea, Hong Kong, China and Vietnam. They needed it to cool facilities that would otherwise have had to cut back production.

So this is what it could come to. The importing of water, to allow the production of oil, to allow further burning of oil, to cause more drought, to force the importing of more water, to allow the production of more oil . . .

I was in Tokyo to do the same thing I had done now in London, Washington, Brussels, Madrid, Zurich, Vienna, Venice, Auckland and Brisbane: to talk about the impact of climate change. I was becoming sick of the sound of my own voice. In the press club of the Japanese Government's Environment Agency, I ran through the story once more, this time in slow motion.

The first questioner asked the obvious. Was this dreadful summer heat because of the greenhouse effect?

I said the same thing I had in London, Washington, Brussels, Madrid, Zurich, Vienna, Venice, Auckland and Brisbane. Not by itself necessarily, but taken with all the other worsts-of-the-century, we in Greenpeace believed so, even though nobody could yet offer cast-iron proof.

I was beginning to wonder if it was the right line to take. I knew from my electronic bulletin boards, and the faxes from home, that the whole of Eurasia was sweltering. 'THE WORLD IS HOT, THE WORLD IS BOTHERED,' a headline in the *Independent* read. 'Mon Dieu, what a scorcher. Britain is not alone in enjoying, or suffering, a blisteringly hot summer this year. Across the northern hemisphere, from France to Japan, weather records are melting (along with roads and, in the Czech Republic, railway lines).' It was true. In temperatures of 29°C, Czech railway tracks had twisted out of shape.

While I was in Tokyo, the thermometer hit 39°C, the city's highest ever temperature. Tens of thousands of chickens died of heat stroke. Scuttling between air-conditioned buildings, and downing gallons of chilled beer, I tried not to join them.

On this visit, Yasuko Matsumoto had also organized meetings with Japanese insurers. I had talked to her Tokyo insurance contacts on the phone ahead of the meetings, saying that I had a bit of a problem. Journalists would think me inconsistent if I didn't talk to insurance-industry representatives while I was in Japan. But if I said I was indeed talking to insurers, the journalists would want to know who I was talking to. What should I say?

You can say you spoke to insurers, they told me, but on no account can you say who, or where. Greenpeace had still not shrugged off the anti-whaling stigma in Japan, it seemed. So I went to my meetings cloaked not only in sweat, but in secrecy too.

At the first meeting our most senior contact told us that property-catastrophe losses in Japan in recent times had this year forced the industry's rating association to ask the Ministry of Finance for permission to raise premiums. This was the first time in forty years that such a request had been made. But it was a heavily political issue, he said. The government deferred any decision. Insurance companies might be able to raise their rates elsewhere, but not in Japan.

'If the situation is as you say,' our contact told me, 'and weather losses increase, we cannot survive. We can't increase the rate, can't increase deductibles, and cannot refuse our customers. What can we do?'

The executive looked at me, and laughed.

I liked to think I knew Japan and the Japanese quite well. But faced with this gallows humour, I just didn't know what was appropriate. I looked at Yasuko. She was laughing too, but in a reserved sort of way. I tried to copy the mood.

'Isn't it possible for you to join forces with the Environment Agency,' I asked, 'where I know there are officials who understand the threats you face? Then you would be able to reinforce each other's arguments.'

'The Environment Agency people are very friendly to us, because they are concerned about climate change. But it is the Ministry of Finance which controls the rating, and they have a very different attitude. They say that because we are in big buildings, we can afford to pay.'

The executive unrolled some colourful maps. He wanted me to understand the flooding risk in Tokyo. The maps showed vertically exaggerated cross-sections across the city and the Tone River, the major river in the Tokyo area. Vast swathes of the city, it was instantly clear, were below sea level. The Tone, in contrast, flowed along what seemed like a viaduct. The walls of this viaduct were the levees of the river. The bottom of the river was much higher than the majority of the city. Small inset photos of buildings around the sides of the map showed what would happen if the river burst its banks in certain places as a result of the torrential rains that accompanied typhoons. Red lines on the photos showed where the water level would rise to. One of these photos showed a tower block. The red line wrapped around it on the fourth floor.

'We still can't reach an accurate estimate of the potential maximum loss if a bad typhoon hits Tokyo,' the executive told me. 'Europeans and Americans come here with simple computer models and think they can tell us our potential maximum loss. We say, "you must be some kind of genius".'

I knew some of the people he probably meant. Their estimates approached $100 billion. I looked out of the top floor window of the big building I was in, across the skyline of the commercial district, and beyond to the infinite cityscape stretching off in every direction to the horizon. I tried to picture the damage a super-typhoon's winds could do, and then on top of that the flooding.

If Japan suffered this way, so would other countries. Capital flowing out of Tokyo in tens of billions of dollars to be invested in other economies would suddenly go into reverse. It would be flowing backwards for the rebuilding process.

AUGUST 1994, GENEVA

As the tenth session of the climate negotiations opened during August 1994 in Geneva, all eyes were on the Germans. They were to host the first Conference of Parties of the climate convention, otherwise known as the Berlin Climate Summit, in March 1995. They held the presidency of the European Union, and their environment minister was set on seeing a protocol to the convention, with targets and timetables for emissions reductions, signed by the end of the Berlin summit. It didn't seem too unrealistic a goal. At the end of the previous session of talks, in February, the industrialized nations had taken a crucial step, albeit on a psychological rather than a practical level: they had collectively agreed that their commitment to freeze greenhouse-gas emissions by the year 2000 would be inadequate. At some stage, they conceded, they would have to begin making cuts.

The carbon-club lobby was the biggest ever seen at the negotiations. The Global Climate Coalition and the Edison Electric Institute each had nine representatives, and the International Chamber of Commerce no less than sixteen. The International Petroleum Industry Environment Conservation Association sent six, the World Coal Institute four, and Don Pearlman's Global Climate Council three. Scattered among this brigade were representatives from Shell, BP, Exxon and Texaco, and they were just the ones whose company affiliation I knew. There were also over twenty chemical-industry lobbyists, there to try to stop the negotiators agreeing any limits on HFCs and HCFCs.

As delegates walked the corridors of the Palais des Nations, so the carbon club's foot soldiers would press the latest thoughts on the defence of oil and coal into their hands. 'To date,' read a highlighted quote from John Schlaes in the Global Climate Coalition's material, 'science has been unable to establish what qualifies as a dangerous level of greenhouse-gas concentrations. This makes a judgement on the adequacy of commitments logically impossible.'

Schlaes had presented his organization's views to the State Department ahead of the session. 'The consideration of common regulatory or financial measures to reduce emissions is inappropriate,' he had argued, 'and must be resisted as US domestic policy or in international forums. The potential damage to the US economy and to its international trade competitiveness, with attendant job losses, just cannot be justified on the basis of the current state of the science.'

The insurers had yet to bite the bullet and send representatives in person to the climate negotiations. So I decided to bring top insurers to Geneva on celluloid. On the first day of the talks, I screened the video made by my colleagues in Greenpeace USA.

Although the screening had to be held during the lunch hour of the first day, usually enough to guarantee a mere handful of participants, over seventy delegates turned up. When the eight-minute video ended, I stood to answer their questions.

Among the seventy delegates were at least half a dozen of the carbon club's hit men. J. R. Spradley was one of them, on this occasion representing the Edison Electric Institute. He had his question ready. There had always been hurricanes, he said. People were just moving into the way of them in greater numbers and betting their house that a hurricane wouldn't hit. I told him to listen to the two biggest reinsurance companies in the world. Their meteorologists needed no further persuasion that global warming had had a hand in the damage bills of recent years.

'And anyway,' I said, 'the more important point is about the future. You and your colleagues are prepared to bet the house – I mean the global house – that there won't be an increase in climate catastrophes. And you know full well how big are the stakes on that bet, and how short are the odds you get from most climate scientists.'

Afterwards, a stream of delegates asked for copies of the video. Among them was an official from the German Federal Environment Agency. He wanted his colleague from the Finance Ministry to watch it.

*

That evening, the Swiss delegation hosted a reception. Once again I stood on the roof terrace of the Palais des Nations, glass in hand, looking out at the beautiful view across Geneva and Lac Leman to the Alps.

My first conversation of the evening was with Exxon's Brian Flannery. We soon got into the old discussion, the one I had had with all of the carbon club's foot soldiers at some point or other: whether or not their assumption that climate change was an overhyped non-problem was correct. But I had never had the discussion face to face with Brian Flannery before. I found him to be even more of a hardliner than I had imagined. He assured me that most of the recoverable oil and gas would be burnt, and that it would have no noticeable effect on the climate.

I drifted on among the talking delegates, reflecting on how Flannery could possibly believe what he did. I wondered how many of the others in the carbon club maintained that belief genuinely, and how they were able to sustain it.

Within ten minutes I had found an answer to the question, at least in one case. John Schiller of the Ford Motor Company was the next person I came across standing without anyone to talk to. Schiller's namecard revealed him to be Emissions Planning Associate with Ford's Emissions Control Planning staff in Michigan. I had never spoken to him at the negotiations before this day, though I knew him by sight as a regular among the industry lobbyists. He had actually come up to me after the insurance video presentation, and I had been mildly heartened by his comments. He had told me that he had appreciated my honesty in dealing with the uncertainties, and that he wanted to know more about the issue. If there was a real problem with the financial markets, he needed to tell his company about it. They weren't just interested, he said, in hearing arguments that climate change was unproblematic, the way I might think.

Now, on the terrace of the Palais des Nations, I led our conversation straight to the question of how he was able to think there was no problem with a world of a billion cars intent on burning all the oil and gas available on the planet.

The first thing to be aware of, Schiller told me, was that the view environmentalists held that all the carbon in fossil fuels had been taken out of the atmosphere over many millions of years was in fact wrong. The world was much younger than that. The carbon had been taken down quite quickly, and therefore it wouldn't matter too much if it was put back up there quickly.

How young is young? I asked.

Maybe 10,000 years, Schiller replied.

This would be news to radiometric daters, astronomers, archaeologists, palaeontologists and stratigraphers. Not to mention the geological exploration and research teams of the oil companies that filled Schiller's Fords with gasoline.

Didn't he have a problem with the fact that the ice cores alone go back 160,000 years?

Oh no, they were dated all wrong. He went on to tell me that there was too much carbon-14, the radioactive isotope of carbon used in radiocarbon dating. He had done a lot of reading on all this, he told me. There was all sorts of corroborative evidence. And then we came to it. 'You know, the more I look, the more it is just as it says in the Bible.'

I told Schiller that I had spiritual leanings myself, and that I would be very interested in hearing exactly what he meant.

And off he went. It was all in the Book of Daniel. The prophecies actually made it clear that it really didn't matter too much about climate change anyway, because the Good Book told us that there would inevitably be growing devastation on the planet. And yes, ecological degradation would be a part of that – I and my colleagues would find ourselves right on that one. It would all lead inexorably to world government.

I listened inertly, making appropriate comments about how interesting it all was. Schiller warmed to his story.

There was bad news and good news, it seemed, in the world government business. The bad news was that the world government would be led by the Antichrist. The good news was that it wouldn't last for long, and idyllic things lay beyond.

The World Council of Churches, I recalled, had told climate negotiators at the UN in February 1994 that accelerated climate change represented not only a threat to life but also an inescapable issue of justice, a view they had offered in a report entitled 'Accelerated Climate Change – Sign of Peril, Test of Faith'.

Test of faith.

'Have you checked with the World Council of Churches how they feel about your thesis?' I asked Schiller.

'Oh, I don't regard them as a Christian organization.'

God, Christ, Allah and Buddah – help!

So where did environmentalists fit into all this?

Schiller's voice assumed an evasive tone for the first time. But the message came through to me just the same. Environmentalists, it seemed, were in league with the Antichrist, whether wittingly or not. We were doing his work for him.

At this session of the climate talks, the Antichrist also had some disciples who were neither environmentalists, nor UN bureaucrats covertly pushing for world government. On 24 August 1994, after five years of climate negotiations at which the carbon-fuel businesses had been represented from day one, the renewable-energy industry made its first appearance.

As the third day of the session closed, delegates and NGOs alike flooded from the negotiating chamber to a nearby hall, where the newly created Business Council for a Sustainable Energy Future was to hold its first meeting. As I walked in, I picked up the new organization's pamphlet, and read the front cover with a smile on my face.

'The Business Council for a Sustainable Energy Future is comprised of environmentally responsible companies that support the objectives of the Framework Convention on Climate Change, including the long-term goal of stabilising the concentration of greenhouse gases in the atmosphere.'

The *concentration*. Meaning deep cuts in emissions.

'The Business Council believes that delay in taking actions to mitigate global warming is irresponsible, and that the opportunity cost of delay is substantial.'

Over twenty companies and organizations were listed as members of the new business council, and they included energy-efficiency and renewables organizations like the Solar Energy Industries Association, the United States Export Council for Renewable Energy, the American Wind Energy Association, Energy Conversion Devices, Inc., and the North American Insulation Manufacturers Association. Gas companies were also mixed in among the names. One of them was Enron, America's largest gas company. As the *Oil and Gas Journal* had feared, civil war had broken out in the oil and gas industry.

The executive director of the BCSEF and three executives from member companies sat facing the hall, ready to address the meeting. One of them was Enron's head of international affairs. I looked around the packed room. Seated among the rows of curious diplomats and environmentalists were some of the carbon club's main players. No smiles here.

One of the executives who spoke was a vice-president of energy conversion devices, Nancy Bacon, representing the US Export Council for Renewable Energy. With high-volume production plants for solar photovoltaics, she said, her industry could provide energy at rates competitive with fossil fuels. 'We have technologies today that can be used for many applications in many countries.'

I allowed myself to enjoy one more glance at the expressions on the faces of Schlaes and his cohorts. For so long at these negotiations they had described themselves as representatives of 'the business community', as though the entire clean-energy sector did not exist, much less the financial sector. Those days, it seemed, were over.

SEPTEMBER 1994, GENEVA

When next I was in Geneva, Don Pearlman was also in town, as usual, but this time we were at different events. Pearlman was at a meeting of the IPCC policy-responses working group, trying to torpedo anything progressive in their contribution to the IPCC's Second Assessment. I was at a seminar for bankers.

At breakfast in my hotel on the first morning, I met an IPCC official. He told me grimly how Pearlman and the other carbon-club lobbyists had recently tried to neutralize a meeting of the IPCC scientific working group, and very nearly succeeded. The IPCC secretariat, the official said, was getting very fed up with Mr Pearlman and the shameless way he used the Saudi Arabian delegation as a proxy for his stalling tactics.

I listened unhappily. I had stopped going to IPCC meetings, so as to concentrate full-time on my financial-sector work. But the IPCC man told me that invariably no environmental NGO representatives turned up at IPCC meetings these days, and that Pearlman and the other oil- and coal-industry people, working with their OPEC proxies, were having an unopposed run at watering down the science and the policy-response recommendations.

If he meant to make me feel guilty, he succeeded more than he knew: my colleague Bill Hare was about to re-enter the fray, if not recovered from his stroke then certainly having taken the decision to ignore it. He had volunteered to cover the IPCC for us until the all-important Second Assessment report, due out late in 1995, was completed. Bill was widely regarded – in government and non-government circles alike – as the best climate lobbyist in the business. No one in Greenpeace's management team was about to instruct him to take more time off, it seemed. If the carbon club succeeded in throwing their spanners into the IPCC works, then the steam would be taken right out of the climate negotiations for the next few years at least. Just a few lobbyists – Pearlman, Schlaes, Flannery and their friends – would have bought their masters in big oil and coal another few years of unconstrained fossil-fuel burning.

I made my way to Geneva's International Convention Centre. A UNEP official, handing me my registration documents, told me that there had been an unexpectedly high level of interest in the meeting. Forty-five of the world's major banks had sent representatives. Along with other invitees from the financial world, some ninety people were now registered.

The room was already decked out UN-style with name-flags along the rows of seats. The Bank of America, Deutsche Bank, the Hongkong and Shanghai Bank, National Westminster Bank, the Union Bank of Switzerland . . . this was going to be interesting. Only three environmentalists had been invited.

I took my place, and saw that the director general of UNEP herself, Elizabeth Dowdeswell, was to chair the meeting.

Liabilities associated with toxic chemicals and contaminated land dominated the early discussion. I waited for climate change to make its first appearance in the programme, which it did after the first coffee break in the form of a presentation by Andrew Dlugolecki of General Accident, the lone insurer at the meeting.

'I'm glad to be here,' he said. 'I don't think there has been enough communication between the different sectors of the financial institutions. I hope to awaken your interest in climate change.'

Dlugolecki explained that he was chair of the IPCC's working group on the financial services industry, and went on to give the same crisp synopsis of the principle of an enhanced greenhouse effect, and the problems it could pose, that I had by now heard him give several times elsewhere. He made the crucial point about the varying timescale of threats to financial institutions: how insurers – big as their climate-related problems were – put their money up only for periods of one year, essentially, whereas banks provided loans over twenty-year periods. 'We in the insurance industry can react fairly quickly,' he emphasized. 'You are trapped.'

No fewer than 22 banks had each loaned more than a billion dollars to the energy sector in 1993, almost all of it for fossil fuels. Loans to oil and gas from the Swiss Bank Corporation, for example, totalled $3,019 million, from National Westminster $1,986 million, from the Union Bank of Switzerland $1,796 million, from the Bank of America $1,133 million, from Deutsche Bank $699 million. This, of course, was only the loans made to entities which were deepening the climate crisis, and threatened by panic policymaking once the world woke up to the crisis. On top of that, in terms of banks' assets at risk, came loans to entities directly threatened by the crisis, such as coastal developers or insurance companies.

When it came to recommendations, Dlugolecki did not limit himself to risk management. The question of investment strategy, he said, needed to be looked at carefully. 'It's a source of continual surprise to me that my company has tens of billions invested around the world, yet the fund managers are not considering climate, and are locking money into investments for twenty years.'

Immediately Andrew Dlugolecki had finished, it was my turn. I had had an unbelievable piece of luck. I had an audience of over fifty top bankers, and I didn't even need to convince them that climate change was a threat to their interests. I could concentrate on spelling out the case that the bankers, like the insurers, must start tackling the risk at source.

In November, the Delphi International group of financial consultants would be releasing an important report showing how the capital markets were overlooking climate change, and doing so at their great peril. The argument was simple. I knew it well, because I had commissioned the report. For example, investing today in a coal-fired power station, expecting it to be profitable for the 30 years of its life, was to completely overlook what climate scientists in NASA and such places were saying about climate change in that kind of time frame. The same applied if you were investing in an oil exploration project, expecting the oil to be just as profitable when it came to market in 15 years' time as it was today. You were making unreasonable assumptions about the basis for the future profitability of fossil fuels.

The Delphi Report, written by a former director of Chase Investment Bank, Mark Mansley, argued that it was only a matter of time before international concern about global warming would swell, and governments would be compelled to press the response button: for example, to bring in carbon taxes and active measures to promote solar energy at the expense of carbon fuels. The Delphi Report argued that the risk to investors was significant. Bad as all this was, there was more. The report made it clear that there were also non-business risks which investors had to take into consideration. As things stood, and as my audience knew well, burgeoning toxic liability costs were not being incorporated into the balance sheets of carbon-fuel companies, even though the evidence of their potential enormity had been there for all to see in recent years. But over and above this, the Delphi Report raised the prospect that as climate-related damage bills clocked up, and the evidence that greenhouse gases are altering the climate became clearer, the world might start to look back in anger. In such a climate, class actions by victims might be launched against those who facilitated acts of greenhouse-gas profligacy in the years after the warning to governments issued in 1990 in the IPCC's First Scientific Assessment.

I explained that the report likened this prospect to the experience of the tobacco industry, currently facing the largest class actions in legal history. The structure of these cases is not dissimilar to the potential cases the oil and gas industry could face, Mansley argued in the report. As he put it, 'allegations of anti-trust activity and consumer fraud in the case of the tobacco industry could find a parallel in the lobbying and campaigning activities of the carbon-fuel industry. The consequences of being held liable for some of the damage arising from climate change could be devastating for the carbon-fuel industry. If climate change costs are as large as forecast, having to pay for even a small fraction of this would severely affect the viability of carbon-fuel participants.'

NOVEMBER 1994, CITY OF LONDON

Delivering the annual lecture of the Chartered Insurance Institute Society of Fellows presented me with a new kind of problem. My daughter, Jess, sat in the audience. A few months earlier, newly graduated, she had landed a job in the insurance industry as a trainee regulator of brokers. The Chartered Insurance Institute being the body which examines trainee insurers of all kinds, Jess would be sitting their exams before long.

I had an hour, plenty of time to carefully set out my full argument. It included, of course, the massive threat to jobs in the labour-intensive insurance industry, should climate change indeed bring about a global insurance crash. I thought of my daughter's pride in her new job as I spoke, how her confidence was visibly growing by the week as she found herself coping with new responsibilities. I was more than usually aware of the need to mix all the gloom with an upbeat scenario, so I dwelt on my favoured prescription for how the new solar era could in principle emerge quicker than people might imagine, and how if it did a host of advantages – social and economic – would come with it.

Then I faced questions. They came and went. I forget what they were. I had answered them all dozens of times before.

A thoughtful-looking man got to his feet in the front row. I had met him already, in the president's room as I drank tea with the

Chartered Insurance Institute's officials just before the lecture. He was the director responsible for examinations, and he was seated right next to my daughter.

'Given all you have told us, and the clear, er, magnitude of the things that need to be done, I wonder, Dr Leggett, er . . . why are you still optimistic?'

Not long after, the City of London staged its first major conference on what the capital markets needed to know about corporate environmental performance. Several hundred city bankers, fund managers, analysts and accountants sat in National Westminster Bank's headquarters listening to a series of presentations on corporate environmental reporting. I sat among these City folk listening to NatWest's chief executive, Derek Wanless, open proceedings by saying that only 110 out of 37,000 transnational corporations bothered to file environmental reports, and that was nowhere near good enough. During the course of the morning, other speakers offering the same view would include the Secretary of State for the Environment and the managing director of BP, from the podium, and a Greenpeace colleague from the floor. With that kind of spectrum of support, the City players must have felt compelled to conclude that it would only be a matter of time before they were dealing routinely with environmental reporting, and factoring environmental performance indices into ratings of equity and debt holdings.

It was the question of greenhouse performance that particularly interested me, and I didn't have long to wait before the issue raised its head. Tessa Tennant, head of ethical investment research at National Provident Institution, told the audience that there was a sea change beginning in the financial institutions, and the driving forces included global threats like climate change. She put over the solution in the form of a challenge. 'I want to put it to you that in forty years' time, when someone asks you, "Were you there during the eco-industrial revolution?", you will be able to say, "Yes, I was a member of the team which persuaded BP to spin off part of their solar energy activities. They made a fortune out of it and they made one of the biggest solar companies in the world".'

Two speakers later, the managing director of BP, Rodney Chase, had the opportunity of responding to that question. Chase had just been appointed chairman of the Business Council for Sustainable Development, now renamed the World Business Council for Sustainable Development after its merger with another group, the World Industry Council for the Environment. It was when Chase had been head of its US arm that BP had begun contributing to the Global Climate Coalition. Here, then, was a man I was bound to be suspicious of: a core funder of the carbon club. No matter what he said in public, I would find myself wondering what words passed his lips in private.

'Those of us who have been given the privilege of developing resources,' Chase began, 'should accept the responsibility for explaining the environmental impacts of our processes. The buck obviously stops with us.' Oil companies had one inalienable imperative, he said: respect for their customer base. And the customer base regarded the environment as an increasingly important issue. 'Society is worried about this. Our shareholders are worried about this. The mailbag that we receive from our shareholders tells me that this issue is very high on their list of interests and concerns.'

Chase had an interesting perspective on how environmental reporting should be achieved. He did not want to see some companies rewarded for saying nothing while others, his included, took the flak for being transparent and forward-looking. 'I suspect most of us feel, rather cynically, that it will take legislative levelling of playing fields and setting of rules for most boards of directors to take the courage to really start to count the cost of environmental management.'

I found myself surprisingly encouraged by this. But of course, the fact remained that Chase had dodged the greenhouse issues completely. In particular, he had side-stepped the specifics of Tessa Tennant's challenge. In the final plenary discussion, I could not resist an attempt to throw her question back at him. Of several dozen raised arms, one the chair pointed at was luckily mine. This morning's *Financial Times* covered a report, I began, which offered the view that enhancing the greenhouse effect threatened the capital markets in a variety of ways. I shamelessly pointed out the piles of Delphi Reports that the National Westminster Bank had allowed me to arrange on one of the display tables at the side of the hall. The core solution in heading off that threat had to be the generation of huge solar markets. So how would Mr Chase answer Ms Tennant's question? In forty years' time, when someone asked him 'Where were you when the eco-industrial revolution took place?', would he be able to say 'Yes, I was there – I was one of the people who persuaded BP to spin off its solar arm, make one of the biggest solar companies in the world, and a fortune?'

BP's managing director walked back to the podium. He wanted to begin with whether or not he believed there was a problem with climate change, he told the audience. It was an extremely difficult area. He was not utterly persuaded that any of us knew the answers. He was persuaded that the *world* saw it as a problem.

'Do I believe that there is an answer with the flotation of a solar company that can in fact turn the world over to a different source of energy that will have demonstrably less impact on the environment? I do not know. I would be foolish to say it would not happen.'

DECEMBER 1994, BERKSHIRE

The day after my encounter with Rodney Chase, his office rang me. Would I please come and speak to BP's 'high flyers' – their middle managers on track for senior positions – at an in-house seminar? They wanted to hear about what I thought their company's agenda should be for the next twenty years.

Were they sure? I asked. The last time the company had invited me to do one of these contact sessions, the *Braer* had hit the rocks.

They said they would take their chances.

I duly arrived at a conference centre in leafy Berkshire so that thirty BP high-fliers could be presented with a free sample of their environmental opposition.

After dinner, I talked to the BP managers under the watchful eye of the head of environmental affairs, Klaus Kohlhaus. To plan for business-as-usual prosecution of their 'core business', as they called it, they needed to assume four things about the years ahead, I said. First, that governments wouldn't speed up climate policymaking. Second, that publics wouldn't grow more concerned, so creating a political currency allowing politicians to compete in finding ways to tackle the greenhouse danger. Third, that the financial sector wouldn't continue its waking-up process, and translate its mounting concern into precautionary action on investments and loans in the energy sector. And finally, they needed to assume that their own constituency – the oil workforce, and even they themselves – could stay blind to what would pretty obviously be growing evidence that climatic disaster was brewing.

Was there any chance that all four conditions could be met? For decades into the future?

No. They had to start the process of repositioning their company, before it was too late.

7

A Mandate Delivered

January–April 1995

FEBRUARY 1995, NEW YORK

The year 1994 had been the fourth hottest since records began, with global average temperatures 0.31°C above the 1951–1980 average. The cooling effect of the 1991 Mount Pinatubo eruption had now well and truly disappeared. The head of NASA's Goddard Space Flight Center, climate scientist Jim Hansen, told the *New York Times* that he was more confident than ever that there was a real long-term warming trend in progress, and that it was due to the greenhouse effect.

February 1995 opened with the worst floods ever in Germany, the Netherlands and France. As though the worst-of-the-century floods of December 1993 had not been bad enough, with $500 million of insured losses and $2 billion of economic losses, the Rhine now rose to fully 17 metres above sea level at Lobith, and with the Lek, Waal and Maas rivers also at record heights, a quarter of a million people were forced to flee their homes in the Netherlands. From the evacuation scenes on television, it looked as though the Third World War had broken out. This was now clearly not a once-in-a-lifetime problem. A Munich Re spokesperson told the press that his company blamed global warming. 'The many warnings made by reinsurers and insurers that the frequency and size of natural catastrophes are mounting are now seen to be right.'

In the front-page press coverage, for the first time in the spectacular post-1987 sequence of billion-dollar climate catastrophes, the words 'global warming' started appearing regularly. Some articles went even further. The *Independent* ran a headline which must have caused a few ripples in the PR departments at BP and Shell: 'HURRICANES FROM THE PETROL PUMPS.'

Some of those thinking of escaping the flooding may have thought about booking a trip to the World Skiing Championships, due to start soon in the Sierra Nevada mountains of Spain. But going there would have been a waste of time. The championships were cancelled for the first time in 64 years, losing investors tens of millions of dollars. The air temperatures were too high. Not even 457 snow-making cannons imported to fight the crisis could make any difference. The only thing going downhill at the beginning of 1995 seemed to be the climate.

On 9 February, Frank Nutter went to the White House to air his growing concerns about global warming to Vice-President Gore. With him went his Reinsurance Association of America board of directors; the presidents of the American Insurance Association, the National Association of Independent Insurers and the National Association of Mutual Insurance Companies; plus the chairman of the huge State Farm insurance company.

I was in New York, at the eleventh session of the climate talks. While I was busy at the photocopier with a copy of the Reinsurance Association's press release, John Schlaes came up behind me. The Global Climate Coalition's chief was in relaxed mood. With him at this session of the talks was his usual huge team of carbon-club lobbyists, and they had already scored a hit. On the eve of the session they had released a study by a weather consultancy, Accu-Weather, which had claimed that there was no convincing observational evidence that extremes of temperature and rainfall were on the rise. Climate scientists at the talks were apoplectic. It was disinformation at its very worst. The temperature claim was based on three supposedly 'representative' stations, all in the USA, and the precipitation claim – incredibly – was based on just one station. There was no mention of the growing data in Europe and Australia showing significant increases in precipitation: 30–40 per cent in Germany during the last forty years, for example. A press conference for the print media had gone very badly, apparently, when the journalists had learned what the analysis was based on, but significant damage had still been done:

the carbon club had notched up some unquestioning television and radio news coverage.

Now Schlaes looked down at what I was doing. He picked up the press release and took in the headline.

'You guys are creative, you know that?' he said slowly. 'You keep us busy.'

John, you monster, I wish I could keep you out of business.

This session of the negotiations was the last before the Berlin Climate Summit, the first annual Conference of Parties to the climate convention. A draft protocol by the Alliance of Small Island States, requiring legally binding cuts in carbon dioxide emissions by the industrialized countries of 20 per cent by 2005, was finally receiving support from some of the big developing countries. Malaysia had broken the hitherto unified opposition among other developing countries to the AOSIS proposals. Now other developing countries were beginning to come out in support of the island nations and their desperate attempt to get the world to agree in Berlin to cuts in greenhouse emissions.

This proliferation of support for a reductions protocol was a setback for the carbon club. If environment ministers went to the Berlin summit with the AOSIS protocol in any way still realistically alive, then in the full glare of publicity at the Summit it might be possible that governments would feel forced to begin negotiations to reduce emissions. They might even feel impelled to agree a reductions protocol before the two-week summit was over.

The carbon club had other troubles. The gas and renewables industries, in the form of the Business Council for a Sustainable Energy Future, had turned up again at the negotiations in order to voice their support for the goals of the convention. The Global Climate Coalition and the Global Climate Council were now caught on the horns of a dilemma, having claimed for so long that the business community had one voice, and that they were it. The two sides of the business divide could not agree how to word the one speech they were allowed to make to the governments. Shouting matches, with Don Pearlman's voice loud among them, were heard coming from the industry NGO room.

No sooner had this music reached my ears, than I ran into one of my contacts on the Swiss delegation.

'Jeremy,' she said excitedly, 'good news on the insurance front. Swiss Re are coming to Berlin. They are going to be represented on our delegation.' Finally, it seemed, the insurance industry was to take its first steps into the arena of the climate talks. This was another encouraging 'first'. At the tenth session of talks, the clean-energy industry had made its first appearance. At the twelfth, in Berlin, the financial-services industry would be making its debut. The voice of industry would be fragmented at last.

MARCH 1995, FRANKFURT

There were times, when I talked about the insurance-crash scenario, that I asked myself whether I was missing something. How had it been possible for an environmental group, with no experience of markets, to find so much mileage in the idea? True, there were plenty of insurance practitioners saying that mass bankruptcy would be the product of a collapse of catastrophe reserves, and that global warming made such a collapse ever more likely. But where were the economic analyses that factored in such a shock scenario? Where were the long feature articles in the financial pages analysing the economic knock-on problems? Where were the assessments of the employment implications of an insurance crash? Sometimes it seemed as though there was an airiness, an intangibility, about the concept. Although nobody had said as much to me, I felt on occasion that those who understood the markets didn't really think such shocks to the financial system could ever happen.

That feeling left me abruptly on 27 February 1995. I awoke to news of the financial collapse of the century. Barings Bank had lost $800 million in serial gambling on the futures markets, seemingly through the unauthorized activities of a single trader in Singapore. The full extent of the exposures was not immediately clear, and the Bank of England had decided not to bail out Britain's oldest bank. Quite apart from the stunning news of the loss itself, the fact that the lender of last resort had walked away in this manner was nothing short of seismic. The world now had to accept, presumably, that the other big finance houses in the City of London – S. G. Warburg, Kleinwort Benson, Schroders and the others – did not necessarily enjoy the ultimate backing of the Bank of England either. The whole episode was likely to prove, as *The Times* thundered, 'a milestone in the secular decline of Britain's financial-services industry'.

The central question, addressed by commentator after commentator across the world, was the universal one. How could it have happened?

On 3 March I flew to Frankfurt, still not having heard a single person pretend to have an answer to that question. Earlier that day, Nick Leeson, the rogue Barings trader who, it seemed, had brought down a centuries-old bank single-handedly, had been apprehended at the same airport by German police. For my part I cleared customs untroubled, even though my appointment in the city also involved trying to embarrass a bank. My German colleagues had invited me to take part in the press conference where they would attack Deutsche Bank as a top funder of climate chaos, and unveil their poster campaign against the bank. The message on the press release, which was already on the wires, was that Germany's biggest bank, in funding 20 billion marks' worth of coal mining and oil exploration, was contributing among other things to the threat of an insurance crash, and hence a disastrous shock to the world's financial system.

Not that Deutsche Bank was alone in this, of course.

Vying with the Barings scandal for column-inches at the time was a huge iceberg. Some 500 cubic kilometres of the Larsen Ice Shelf had broken away from Antarctica and was drifting out into the Weddell Sea. The phenomenon had first been spotted on satellite photos, which were now appearing in newspapers the world over. The British Antarctic Survey had dispatched a plane to see what was going on.

'From the aircraft window I was utterly amazed to see the dramatic changes,' said one scientist. 'In twenty-five years of Antarctic fieldwork I have never seen anything like it.' The ice shelf in question had spanned 2,000 square kilometres when he began that fieldwork, and scientists with dog teams had crossed it regularly. Now, after this latest and most spectacular collapse, it occupied just a quarter of that area, and was a dangerland of fissures.

'There is very little doubt in my mind that all this is climate-related,' British Antarctic Survey glaciologist David Vaughan told the press. The average air temperature over the Larsen Ice Shelf had risen by 2.5°C since the 1940s, the greatest change in surface temperature anywhere on the planet.

With the media coverage of the iceberg at its height, scientists at Germany's main climate laboratory, the Max Planck Institute for Meteorology in Hamburg, issued a report concluding that their latest studies of the increase in global average temperature since 1880 had left them 95 per cent confident that the increase was not due to natural variability.

Within days of the iceberg reports, researchers at the Scripps Institution of Oceanography published results of a survey of marine life in an area of some 130,000 square kilometres off southern California, where the surface water had warmed by 1.5°C since 1951. The warmer water had become less well mixed over the period, and consequently the biomass of zooplankton has decreased by fully 80 per cent. Associated with this decline had been a drop in fish and bird populations, including a 60 per cent decline since 1986 in the population of the sooty shearwater, a bird that eats plankton and plankton-feeding fish. Dean Roemmich and John McGowen, the researchers who uncovered this depressing tale of biological impoverishment, concluded that the suppression of nutrient supply by enhanced stratification was not a mechanism confined to coastal oceans. If there were to be a global temperature rise of 1 to 2°C in the next forty years and stratification increased globally, the biological impact could be devastating. The IPCC had shown that a 1 to 2°C temperature rise is a virtual certainty, if not in the next 40 years then sometime beyond that – unless we cut greenhouse-gas emissions significantly.

'If it's due to the greenhouse effect,' McGowen told the press, 'it will only get worse and we are in deep trouble.' He thought that the die-off may already have extended across a much larger area. He pointed to evidence of warming across two and a half million square kilometres of ocean in the north-west Pacific, and observed that nobody knew what had been happening to the plankton because nobody had been conducting any measurements.

In the immediate countdown to Berlin, the science of global warming was hotting up in a depressing way.

MARCH–APRIL 1995, BERLIN

In the foyer of the Berlin Congress Centre, two days before the climate summit was due to begin, I felt the churning trepidation of all who take the risk of organizing events, whether they are children's birthday parties, car boot sales, or seminars to discuss the impact of climate change on the capital markets. At this day-long event, to my great good fortune, top insurers and bankers had agreed to speak. What I needed now was for an audience of their peers, not to mention the world's press, to turn up. I had been disappointed in these aspirations before.

I had not been to Berlin since the evening in 1989 when I had crossed the Wall, a month before it came down. So much had changed. Clubby bistros and music bars were now to be found in the streets of the former East Berlin. A hall built into an arch of a railway bridge near my hotel, once a place in which East German border guards were briefed before they headed off to their machine-gun posts, was now an art deco restaurant. The cellar of the house in which Bertolt Brecht had lived was also a restaurant, and where Germany's famous dramatist had once rifled through his wine rack, I had the night before my seminar dined with the half-dozen bankers and insurers who had agreed to stick their necks out for me today by speaking at the event. After the food, while briefly running through the batting order for the event, I had risked telling them my favourite Brecht quotation. 'The chief aim of science is not to open the door to infinite wisdom, but to put some limit to infinite error.' I had no idea what the original context was, I said, but to me that was what global warming was all about. It was what I wanted our seminar to focus on: translating the scientific threat assessment into a threat assessment for the financial sector.

As a queue began to build at the registration desk, I began to relax. I watched representatives of some of Europe's largest banks and insurance companies sign in to listen to the speakers and take part in the day's discussions. The journalists among them represented some of the blue chips of that profession too, including the *Financial Times*, *Newsweek*, *Der Spiegel*, Reuters and Associated Press. Several television crews began setting up their cameras at the back of the hall.

I was secretly amazed that my guests had agreed to my inviting the media into the seminar. Included on the platform that day would be Rolf Gerling, chairman of the giant Gerling insurance group in Germany; Andrew Dlugolecki, assistant general manager at General Accident; Frank Nutter, President of the Reinsurance Association of America; Sven Hansen, head of environmental management at the Union Bank of Switzerland; Hilary Thompson, head of environmental management at National Westminster Bank; and Tessa Tennant, head of ethical investment at National Provident Institution. I had decided to go for broke, and had told the speakers at the time I invited them that what I really wanted to see, apart from a good day's discussion, was media coverage all over the world: I wanted them to help me stage an event, on the eve of the Berlin Climate Summit, which would put out the simple message that top insurers and bankers thought global warming was a grave threat to economies. Too many people, and too many governments, still thought that even if global warming was real, its impacts would be limited to the developing world, and to coasts and islands at that. The speakers had all agreed.

With participants still filing in to take their seats, Rolf Gerling took to the podium to begin proceedings. The billionaire chairman of the insurance group which bore his name was also chairman of this event. Gerling, a philosopher, was a man much concerned by the state of the environment. I had met him some months before in Cologne, when I was invited to his home for an audience along with Greenpeace International's executive director. He was the reason why Germany's TV stations had turned up: Gerling and Greenpeace – a mix strange enough to make the national news. The Greenpeace Germany press office told me that the event had attracted maximum press interest: journalists were fascinated to see what line he would take. The more so because a major shareholder in Gerling's group was a certain Deutsche Bank.

Greenpeace's campaign against the bank had fallen completely flat.

The bank had simply ignored us, and the issue had failed to resonate with enough members of the German public to generate significant pressure. My contacts within the bank told me that senior management was incandescent with anger that we had attempted to mount a campaign against them, and I was filled with admiration for Gerling for risking an appearance at a Greenpeace seminar in the immediate aftermath of this.

This admiration was to swell as the day went on. Gerling had some preliminary remarks to make before introducing the first speaker. 'Our gathering here today is a somewhat historic one,' he said. 'To my knowledge, this is the first time that representatives of the insurance, reinsurance, banking and pension-fund industries have gathered together to discuss the issue of climate change, and the threat that human-enhancement of the greenhouse effect poses for the financial sector. That we do so in the company of representatives of the environmental groups, and the solar-energy industry, adds a further dimension of novelty – and perhaps risk – to the occasion.'

Why are we here, he asked, this seemingly incongruous group? 'We are here because most of us – perhaps all of us – are worried about global warming, and the economic disruptions that may accompany it. We are not alone in this concern, of course. On Tuesday, over a hundred and fifty governments will convene not far from here to continue their negotiations under the Framework Convention on Climate Change.'

Andrew Dlugolecki spoke next. 'Looking forward,' General Accident's climate-change expert concluded, 'I am sure climate change will speed up, and I'm sure that will have major implications for the industry.'

Then came Frank Nutter, the only American among the speakers. Nutter was under pressure. Soon after the session of the climate talks at which John Schlaes had seen me distribute the Reinsurance Association of America's press release about their White House meeting, Nutter had been called to the Washington headquarters of the Global Climate Coalition for consultations. He told me that on the appointed day he had been confronted by a committee of carbon-club heavyweights, with Schlaes in the chair, and a scientist – he could not remember the name – waving thick reports containing the definitive proof as to why he, Frank Nutter, had been misled into thinking there was anything to worry about with global warming. Nutter had elected, notwithstanding, to speak at the Greenpeace seminar in Berlin. Whereupon Schlaes had left a message at his hotel, to await his arrival in the city, to say that he too would be at the seminar, and listening to every word Nutter had to say.

Frank Nutter had simply ignored this crude attempt at intimidation.

'The uncertainties associated with climate change are probably the insurance industry's best reason for taking this seriously,' Frank Nutter concluded. 'Doing nothing is probably not an option.'

Gerling called for questions at the end of Nutter's talk. The first was to the point. Did Mr Nutter know of any insurers who were, as a result of the concerns he had aired, switching investments away from fossil-fuel companies to greenhouse-friendly companies?

'In the US,' Nutter replied, 'I have not seen a change in investment practices. But I hope that happens.'

In the coffee break, immediately after Frank Nutter's talk, I saw John Schlaes coming towards me through the crowd, smiling. He now shook my hand. 'A very interesting event, Jeremy. As always.'

Schlaes did not stay for any other talks. The Global Climate Coalition chief had been true to his promise to try to put the frighteners on the Vietnam veteran who headed the Reinsurance Association of America. He had failed.

At the summit the next day, his organization put out its first batch of publicity. The Global Climate Coalition, it claimed, spoke for over fifty-five trade associations and companies 'representing virtually all sectors of United States industry.' Prominent on the Coalition's delegation, however, were representatives of Exxon, Mobil, Chevron, the American Petroleum Institute and the American Automobile Manufacturers. I didn't notice any from the insurance or banking industries.

The Berlin Climate Summit opened two days later in the city's second international conference centre, a building the size of an ocean liner which looked like a Spielberg starship from the outside, and an amalgam of airport terminals on the inside. Thousands thronged the halls and corridors, mingling among exhibits mounted by every kind of interested party. Outside, a 'monk' began a vigil which would last throughout the two weeks. He held a simple banner exhorting delegates to support the AOSIS protocol. Atop the 190-metre smoke-stack of one of Germany's coal-fired power plants, near Cologne, Greenpeace climbers conducted their own version of that same demonstration. They too were there for the duration, with their banner urging the world to 'Stop CO_2. Go Solar'.

With the first day of the summit came a major change for the carbon club. It had been five years ago, at the World Climate Conference, that the world's press had last taken such an interest in global warming, and now in the immediate absence of exciting action in the negotiating chamber, hundreds of journalists were tramping the corridors in search of stories. Among them was a TV crew from Channel 4 News in the UK. They asked a few questions about who were the key power-brokers in the negotiations, and they were pointed in the direction of Don Pearlman and John Schlaes. Pearlman had enjoyed four years of virtually unrestricted access to the Saudi and Kuwaiti delegations at the talks. Nobody had ever hindered his transparent interactions with them, as he issued instructions and conducted tutorials over text. He, like us, had come to take it all for granted. Now he was being asked to give a television interview in which he would have to explain himself. He refused. The TV camera turned on him anyway, following him around as he patrolled up and down outside the negotiating chamber, waiting for his OPEC henchmen. He suddenly found his ability to operate impaired. Worse, the TV reporter then tried to interview him against his will. TV reporters called this tactic 'doorstopping', I learned. Channel 4's man, thrusting a microphone at Pearlman, asked him whether it was true that he tried to manipulate and stall the negotiations using the Saudi and Kuwaiti delegates.

Pearlman, looking shifty, trying to escape, offered no comment. It would be the first of several such encounters for Don Pearlman with TV crews from different countries. This, the first of them, would even be shown on CNN.

On the first evening of the summit, the city of Berlin held the biggest reception I had ever been invited to. I wandered, glass in hand, past many hundreds of people without recognizing a single face. Then, in the gloom of the warehouse-like hall where the event was being staged, I came across Don Pearlman and John Schlaes.

I walked up to them. I told them that two visiting business delegations had asked me to arrange meetings for them during the summit, one from Lloyd's of London, and one from the British Bankers Association. I wondered if the Global Climate Coalition might like to meet them.

Schlaes had coal in his eyes as he listened to this.

'We can discuss that tomorrow,' Schlaes drawled. In his manner was something new. The condescension, the jocularity, had gone. Beside him, Pearlman was the same.

J. R. Spradley was at the reception too. He was at the summit on the Edison Electric Institute's delegation. This time we were joined in our sparring by someone I very much wanted J. R. to meet: James Anderson, an insurance broker from Lloyd's of London. Anderson had come to my eve-of-summit seminar, and was now preparing the ground for the fact-finding visit to Berlin of Richard Keeling, former deputy chairman of Lloyd's.

I had told Anderson all about the carbon club's key players, and he soon heard for himself Spradley's special brand of nonchalance about global warming. J. R., loosened by a few glasses of beer, was as devil-may-care as always. As usual, his line of argument seemed more to follow the impossibility, in his opinion, of slowing the burning of fossil fuels, than any serious questioning of the science underpinning my threat assessment.

'You know,' Anderson told him, 'the insurance industry could begin to disinvest in oil and coal.'

'That's your choice,' said J. R. 'But I'm tellin' you, you're not going to bully the Seven Sisters.'

The next morning, I travelled across Berlin to another summit being hosted by the city of Berlin. The hundred-plus mayors at the Municipal Leader's Summit on Climate Change considered their event to be just as important as the intergovernmental summit. They came from cities in 65 countries, and represented 250 million people. By the year 2000, half the world's population would be living in cities, which would be extremely vulnerable to the impacts of climate change, especially the coastal ones. Over and above these considerations, the mayors felt they had made much more progress than the governments would: later that day they would be signing a declaration urging cities all over the world to make a commitment to cut carbon dioxide emissions by 20 per cent.

Back at the Berlin Climate Summit, I found the Lloyd's of London delegation. Richard Keeling, James Anderson and their colleague David Mann had begun their day by meeting the chairman of the IPCC. Bert Bolin had given them grim news about the latest science, Keeling told me.

Later in the morning, after hearing tales of frustration and procrastination from inside the negotiating chamber from a number of government delegations, the Lloyd's three met the head of the Met Office's climate-modelling centre. It was the first time the UK's premier climate-research body had sent exhibitors to a session of the climate talks, much less people of this rank. There was a reason. The Met Office's latest climate model had convinced, among others, the UK minister for the environment, John Gummer, that humankind had a huge problem. Gummer had sworn to come to Berlin to try to lead a charge for progress. He was a Conservative, he had apparently said, so he wanted to conserve the planet.

We had lunch with Gerhard Berz, Munich Re's technical chief. Berz had been enthused by what he had seen so far in Berlin, and now told us he thought the industry ought to have a permanent presence at the climate talks. Keeling, Mann and Anderson agreed. They would be saying as much in the report they would be circulating in the London insurance market on their return.

Then we were off to talk to the business lobby. Something told me John Schlaes would not be taking me up on my offer, so instead I asked the International Chamber of Commerce for a meeting. This body was supposed to be taking care of the interests of all sectors of the business community when it came to climate change. But heading the ICC's delegation of forty in Berlin was Clem Malin, head of the communications department of Texaco. Shell and Mobil were also prominent in the ICC's delegation.

I did not know Malin. It would to be interesting to see how sophisticated he was. After all, he was supposed to be representing the interests not just of the oil business, but of the insurance, banking, tourism, farming, medical, water and fisheries industries, and others that would be threatened by global warming.

Malin began by telling Keeling that dramatic and decisive early actions to deal with global warming could have negative impacts on economies. Keeling listened, looking doubtful. What were Malin's experts telling him about the time we would have to wait before we needed to act?

'We won't know all we need to know by 2100.' By the time we got to 2200, said Texaco's slow-speaking public-relations chief, it was also possible that we would have had some kind of technical breakthrough which would have made the problem go away. 'Someone discovers the gene that eats the carbon dioxide or something.'

James Anderson was looking down at his notes, trying to hide an incredulous look. I had said nothing throughout, and now was not the time to start.

'The trouble is,' Keeling said calmly, 'that wipes us out.'

At the negotiations the USA was establishing itself as the chief naysayer once again. Key developing countries had now sided with AOSIS, supporting their plea for a 20 per cent reduction target for carbon dioxide emissions. Critically, this list now included India, China, Indonesia, Malaysia and Brazil. But it seemed that the USA was not even interested in a mandate for continuing negotiations with the aim of agreeing a protocol for reducing emissions in two years' time.

Bill Clinton's diplomats were proving to be no different in style or substance than George Bush's had been.

I pressed on with my financial sector work. I organized a further round of meetings for the British Bankers Association's assistant director, Peter Blackman. The British bankers' representative would speak to only one journalist, from the most conservative of the British papers. 'It is clear that climate change will have a dramatic impact on the work and lifestyles of our business and personal customers,' the *Sunday Telegraph* quoted Blackman as saying. 'It could bankrupt some of them, and make some of them homeless and jobless.'

But as usual, such coverage was not the universal rule. The very next day *The Economist* ran a typically quixotic editorial. 'Most actions would pose a bigger threat to human well-being than does global warming,' it scoffed. 'Consider first, the uncertainty of scientists about the extent of global warming. Despite recent advances, science still understands little about the world's climate.'

The 'logic' of these two, adjacent, sentences defeated me. If we understood so little, how could the scoffing editor be so sure that the cost of action would exceed the cost of impact?

But many delegates saw the editorial. The carbon club made sure of that kind of thing.

The weekend came. I went again to Brecht's old house for dinner, and found J. R. Spradley at an adjacent table. He told me he hoped I didn't have any television crews with me. Evidently he had been filmed even at dinner the previous night. I went to a jazz club for lunch the next day, only to find Brian Flannery at a nearby table. Exxon's man liked jazz. Perhaps there was hope after all.

There certainly seemed to be a little hope in the air that day. In the afternoon I walked down Unter den Linden, the main drag of imperial Berlin. Romanesque buildings splashed with the pockmarks of fifty-year-old bombs marched on either side of the wide avenue towards the Brandenburg Gate, and the strip of land where the Berlin Wall had once stood. The air trilled with the sound of thousands of tiny bells. The kilometre-long road had been blocked by a hundred thousand cyclists. The police had given up, and were ambling, lost to one another, among the throngs of protestors. As a huge German flag drifted in the breeze on the Reichstag, a TV camera soared on a crane above the Brandenburg Gate, recording the biggest global-warming demonstration the world had yet seen.

Der Spiegel, the leading German news magazine, had set three reporters the task of unearthing all they could about the Global Climate Council and the Global Climate Coalition. On the Monday morning of the second week of the Berlin Climate Summit, the fruits of their labour appeared in print on the news-stands.

Within an hour, an English translation was in circulation. As I walked around the convention centre, I saw diplomats grinning as they read the script in coffee bars. I picked up a copy. 'HIGH PRIEST

OF THE CARBON CLUB' announced the headline above a photo showing Don Pearlman with the head of the Kuwaiti delegation, Atif Al Juwaili. Al Juwaili had his hand up, trying to block the camera; Pearlman was turning towards the photographer, a scowl on his bulldog face. The reporters had opened their piece with an account of how they had watched Pearlman give out his orders of the day. But Pearlman, unlike his 'permanent accomplice', as the *Der Spiegel* writers referred to Schlaes, kept the clients of his lobby group secret. All anybody knew was that he worked for Washington law firm Patton, Boggs & Blow, well known for its influence in the lobbying world.

The late Tom Brown, who died in a plane crash in the Balkans in 1996, had helped to establish the leading reputation of Patton, Boggs and Blow as super-lobbyists. His clients were reported to have included not only Sony and American Express, but also the Haitian dictator Duvalier. The firm also acted for Abu Dhabi, a majority shareholder in BCCI, the collapsed drug and money-laundering bank. A former employee of the firm had gone on record as saying, 'the biggest compliment you can give any lawyer from Patton is that he'll do anything for money'.

I read on. *Der Spiegel*'s team had built up an impressive case, trapping Pearlman twice. A Dutch climatologist had told them about Pearlman's tampering, via the Kuwaitis, in the IPCC process. At a critical meeting, the Kuwaitis had evidently tried to submit amendments, in Pearlman's own handwriting, of otherwise undisputed statements. The article went on to describe the scene at a vital late-night session of talks in New York in February, where the carbon club had so blatantly ferried instructions to the OPEC delegations that shocked governments had complained. UN officials had then told all lobbyists to quit the negotiating floor. Pearlman refused, until a UN official had threatened to have guards throw him out. Pearlman had denied to *Der Spiegel* that such a thing had happened. A UN official, Jacob Swager, had bravely confirmed it, on the record.

The next day, the Earth Council organized a panel discussion for the business community. In the chair was Maurice Strong, who had moved from being secretary-general of the Earth Summit to chairman of the huge Canadian electric utility Ontario Hydro. Strong had a panel of industrialists and environmentalists to introduce. Each would make a short position statement, and then debate would commence.

Malin's turn arrived. I was fascinated to hear what the Texaco man would say on behalf of the International Chamber of Commerce. Over four hundred people sat listening to him: journalists, delegates and NGOs.

Today, Malin said, as every day, Texaco would be spending $10 million on its capital development, and it represented less than 2 per cent of the energy business. Was there anything to worry about here? No, not given the uncertainties and ignorance about the science. Texaco was providing much needed fuel for economies. 'We are not part of the problem. We are part of the solution.'

Most governments evidently did not agree. We were now into the final three days of the Berlin summit, and the environment ministers had arrived for the endgame. On Wednesday 5 April, Chancellor Kohl gave a speech to the summit. 'We do not have the right to destroy the soil on which the food of our children, and their children, must grow,' said the German leader. Germany was committed to its goal of cutting carbon dioxide emissions by 25 per cent by the year 2005, and the German government believed that it could be done profitably, which would be good for the German economy. 'Let this message go out from Berlin. We are concerned for the future of the earth, and are ready to take ground-breaking decisions.'

Soon after, I watched John Gummer give his speech. Gummer was not an environmental progressive on all issues, but on global warming he was by now one of the most prominent ministerial advocates of action in the world. 'We are talking severe climate change within the working lifetimes of our children,' he emphasized.

The next morning, as delegates milled around awaiting the beginning of the second day, six young 'monks' appeared in the crowd. They wore brown robes, with pointed hoods and rope belts. Their feet were bare. In a line, their hands held in an attitude of prayer, they walked slowly through the throng towards the high priest.

Don Pearlman saw them coming, and walked quickly away. The line of monks followed him. Down an escalator went the high priest of the carbon club. So too did the monks. Back up the stairs next to the escalator came Pearlman. So too did the monks, a press photographer now snapping the scene. For several minutes this procession wove through the crowd of diplomats, lobbyists and journalists. Grins appeared on many watching faces. Then the UN's blue-jacketed security men arrived, and detained the monks.

I retreated to a coffee bar, leaving the negotiations over the status of the monks to some of my colleagues. The World Council of Churches delegation, it seemed, was contemplating an official complaint about the reason given for the ejection of the young protestors – inappropriate attire.

That afternoon, in the main hall, fifty young people stormed the stage and dropped banners from the balconies. Ten minutes of pandemonium ensued, as the UN's outnumbered guards did what they could to restore order amid chanting youngsters exhorting delegates to support the AOSIS protocol. The protestors were too late, of course. Behind closed doors elsewhere, the wording of the Berlin Mandate was in the process of being agreed. This document, effectively the rules of the road for the next phase of the negotiations, talked about the need for reductions. The next head of delegation at the microphone, once the hall had been cleared, was Ambassador Slade of Western Samoa, the elected AOSIS spokesperson. Slade now tried to condense every ounce of emotion that the island nations collectively felt into a five-minute statement. It moved people to tears, I later learned. I missed it. I was trying to fix up a meeting between the insurance industry's representatives and the head of the US delegation.

On the final morning of the Berlin Climate Summit, the text of the Berlin Mandate was released to the world. The nations of the world had agreed that strengthening the commitments by industrialized nations in the Convention on Climate Change was 'the priority' for continuing negotiations, and that in doing so 'quantified reduction objectives' would be set within 'specified time frames'.

The R-word was in there, at least. This may not have been the AOSIS Protocol, signed, sealed and delivered, but in terms of realpolitik it was more than many had expected. After the disappointments of the World Climate Conference and the Earth Summit, it was not quite third time lucky, but it counted as very bad news for the carbon club.

With text in hand, I toured the corridors, offering the obligatory quotes to journalists.

I asked J. R. Spradley in passing for his reaction. 'I think the major carbon criminals of the next century have got off the hook,' he flashed at me. I presumed he meant China.

'Doesn't that make the USA the greatest carbon criminal of this century, J. R.?'

'That's what you say. I hope so.' Clarity to the fore as ever. 'China and India will burn coal come what may.'

I came upon Clem Malin, and asked how he saw it all.

'Puzzled. Puzzled that half of the world can be let off the hook.'

'They'll move. They don't want to commit suicide.'

'Neither do we.'

As we talked, a TV crew began filming us. It was NET, I saw, a conservative American TV station. It would now be my turn to suffer the experience that had earlier befallen Don Pearlman.

I stood my ground, and agreed to an interview with a sharp-mannered young journalist. A million conservative Americans might watch, hating me, but at least I would have a tie on, and wouldn't look as though I was indulging in subversive activities like Pearlman had on ITN and CNN.

After reading in *Der Spiegel* about all the work Pearlman's law firm did for the likes of Guatemala's military, I wasn't about to solicit the opinions of the high priest and his permanent accomplice. I waited for their press release.

'It's clear the agreement gives the developing countries like China, India and Mexico a free ride,' Schlaes offered as his top sound bite. 'And the Berlin Mandate puts US jobs, economic activity and international competitiveness at risk.'

They were moving immediately to attack the soft underbelly of the process, the fear that even if we in the industrialized world acted, it would come to nothing because China and India, with their vast coal deposits, would do nothing. If the USA could be persuaded to make action on their part contingent on action by the developing countries, the anger of the developing countries would know few bounds. This was a dynamic that the carbon club knew gave them a chance one day of tearing the negotiations apart from within. They would emphasize this issue constantly over the two years ahead.

At the foot of the press release the Global Climate Coalition was described as an 'organization of business trade associations and companies'. I looked for the claim that the coalition spoke for 'virtually all sectors of US industry'.

It was no longer there.

APRIL 1995, SAN ANTONIO, TEXAS

The company most responsible for sparking off the greenhouse civil war in the hydrocarbon business had been the American gas giant, Enron. They were the lead corporate player in the Business Council for a Sustainable Energy Future, but their case up to now had been based on extolling the virtues of gas as an alternative to oil and coal in keeping greenhouse-gas emissions down. Immediately after Berlin, they threw another spanner into the carbon club's works. This one was labelled 'solar'.

The week after Berlin finished, I was in San Antonio, Texas. In a hotel conference suite one block from the Alamo, I listened to Bob Kelly, an executive vice-president of Enron, address an audience of over six hundred people at the annual trade fair of the US Solar Energy Industries Association. The Enron man, despite being an oilman, as he put it, saw the greenhouse threat not just as an environmental problem, but as a market-driver. It was inevitable that global warming would one day force solar into energy markets. 'We see a great big market out there, and we are going for it,' he enthused. He went on to describe how his company now intended to build 100 megawatt photovoltaic power stations generating electricity at 5.5 cents a kilowatt-hour, competitive with fossil fuels. They were actively pursuing plans to do so in Nevada, China and India. They had formed a partnership with Amoco, owners of Solarex, the biggest American solar-cell manufacturer. Using Solarex's second-generation cells, of the amorphous silicon thin-film variety, Amoco-Enron Solar intended to build new manufacturing facilities. If they built just one of their 100 megawatt plants, they would be halving the cost of solar cells, and doubling world production. Moreover, they were intent on building not just one, but many, and wherever they found the right conditions to build their power plants they would also be building manufacturing facilities. They would also be selling solar cells for use on rooftops, on both commercial and domestic buildings. Currently there was no mass production, because there were no big markets. Enron had figured out a way to deliver big markets, and so break solar-photovoltaic energy technology out of its current price trap.

I sat listening to Kelly, enthralled. I recalled what a former director of strategic planning at Texas Instruments, Paul Maycock, had told me in Washington back in February. 'If we could cut the price of PV by three times, to 6 cents per kilowatt-hour, then nobody would use anything else for electricity in the sunbelt. Ever.' He had said the scope for solar PV, if the price trap could be broken, was exactly analogous to the photocopier: the first one of those had cost $30,000. Maycock, I knew, was one of Enron's top consultants.

I thought about a lecture I had heard Shell's head of renewable energy supply and marketing, Roger Booth, give at an International Solar Energy Society conference in London earlier in the year. He too had been upbeat. Covering just 0.4 per cent of the land area of the globe with 15 per cent efficient PV cells, Booth had calculated, would supply the equivalent of all the world's current primary energy needs. Desert areas, for example, amounted to 10 per cent of total land area. In theory, just a small fraction of the Sahara, covered with PV, could provide all of Europe's electricity. 'All the world's energy could be achieved by solar many thousands of times over,' Shell's man had stressed.

But Booth had seen it as an unattainable dream. Every second, 40,000 litres of fuel was being pumped into fuel tanks, he had said. We were not going to be able to change that in a hurry. Maybe not. But if one company took the lead in the way Enron seemed to be intent on doing, many previously unrealistic options would open up.

For a week in San Antonio, in a luxury hotel kept permanently air-conditioned by fossil-fuel electricity at a temperature too cold for shirtsleeves, I mixed with the US solar industry and those among the

US utility industry with interests in solar: listening, learning, making contacts. The proximity of the Alamo seemed on occasion all too appropriate. At one meeting, hundreds of solar-industry and electric-utility practitioners listened to a Department of Energy official describe how a single $5 million grant from a federal initiative was to be doled out. This was pathetic: hundreds of people hoping for crumbs from an annual programme amounting to precisely half what Clem Malin boasted Texaco spent on its capital development each day.

But on the other hand, I detected a stubborn optimism in the industry, and its supporters in the electric utilities. One leading light of the US Utility Photovoltaic Group summarized the past. In 1974, they had thought that most houses in North America would have solar PV and solar thermal installed on them within five years. Now, twenty years later, they were saying the same thing. They had to be right sometime soon. 'Dammit,' he said 'it took 22 years to commercialize the television!'

And the public, it seemed, were itching for the solar revolution to take off. Shortly before the trade fair, a Republican pollster had found that two-thirds of Americans wanted solar programmes to be saved at all costs in the upcoming round of budget cutting. This was just one of a number of very encouraging poll results discussed in San Antonio, and utilties' surveys of their own customers' opinions also threw up encouraging news. I was particularly encouraged to hear that in California, the Sacramento Municipal Utility District was in the process of commercializing PV where the economics made no sense. People were paying the utility 15 per cent on their bills to put solar on their roofs, even with no payback for the electricity generated; they were that keen, it seemed, for the utility not to build any more conventional power plants.

The many uncertainties arising from deregulation of the electricity supply, as it stood, were presenting the utilities with severe strategic positioning problems. The shifting sands of consumer desire and potential environmental market drivers added to those problems. The identity crisis in the industry was palpable.

A manager at Solarex was among those who loved what he was seeing and hearing. People would soon be able to festoon their roofs and backyards with solar, he told me. In an age where their consumers could effectively turn their homes into stand-alone power stations, the utilities were going to have to change or die. The waiting markets were huge. Solar PV was selling today at $7 per peak watt. Prices were dropping by 9 per cent a year, even without market accelerators like Enron's scheme. The price would surely be down to $3 per peak watt as early as the year 2000. When that happened, the US markets for solar, both off-grid and on-grid, would be monstrous.

How I wished I could somehow have transplanted the Berlin gathering of insurers and bankers into San Antonio.

APRIL 1995, HOUSTON

I needed to know more about Enron's solar push. It seemed genuine, but I had to be sure. Here was the biggest gas company in the world, with oil interests as well, joining up with Amoco, one of the Seven Sisters, to launch a venture which could provide the catalyst for the take-off of the solar-photovoltaic market. Was there something I was missing?

I flew to Houston the next week to see Bob Kelly in Enron's glittering downtown skyscraper. The co-chairman of Amoco-Enron Solar, as his card announced him to be, gave me an hour and a half, one on one.

Kelly told me how it had all come about. He was an oilman through and through, he said, but two years ago his chief executive had given him the task of looking at energy and the future. What were the new alternative-energy technologies to back? How should Enron position itself? Kelly had looked around, and decided on solar PV.

Kelly, a frank and likeable man of Irish ancestry with a distinguished earlier career in the US Army, told me that not everyone in Enron approved of what he was doing. Kelly's stock was high in the company after he had presided over, among other successes, a huge gas-fired power-plant deal on Teesside in the UK. He had political capital in the company's bank, and now seemed intent on investing it in solar energy. This was the future, he said, and he was going for it not just for business reasons. He was convinced, he said, of the global warming threat. 'I may be muzzled on this, but I'm also a bit independent. I need a logical argument to shut me up.'

I left the meeting with no doubts that Bob Kelly was for real, but less than certain about his power base at Enron.

The carbon politics spawned by this state of affairs were fascinating. Enron helped fund the Business Council for a Sustainable Energy Future. Amoco helped fund the Global Climate Coalition. These two bodies were lined up at the climate talks on opposite sides of the fence, at the level of shouting matches in the UN. How was that going to play out, ultimately, in terms of their new solar partnership? This was far from clear.

Whatever happened, I knew I had to do all I could to foster the new alliance.

APRIL–JUNE 1995, LONDON / AMSTERDAM

If Enron and Amoco could produce this kind of thinking about solar energy and the future, why not BP and Shell? Together with Greenpeace UK's solar-energy campaigner, I had visited BP's solar-energy company at its UK headquarters in March, just before the Berlin Summit. It had been a deeply schizophrenic meeting. The three executives we met at BP Solar that day had been full of tales of the potential size of solar markets, and alive with ideas for how to develop them. I saw nothing to suggest that they were anything other than genuine solar enthusiasts. But BP Oil sent two traditionalists down from their head office in London. They clearly had a brief to pour as much cold water on the proceedings as possible. I exhorted BP Solar's executives to factor into their corporate planning the prospect that public concern about global warming might take off and so drive huge solar markets. The head-office hacks duly argued that I was peddling environmentalists' wishful thinking: non-problems did not drive markets.

As for Shell, the day before their historic U-turn with the *Brent Spar* in June, all my suspicions were confirmed about that company's approach to solar. In Amsterdam, I met a former Shell employee who was very familiar with the company's solar efforts. Clearly a frustrated man, he told me and two of my colleagues a chapter of shameful stories about how Shell's senior management had sat on opportunities, over the years, to expand the solar markets. He showed us a graph which perfectly illustrated the oil giant's cynicism. It plotted the classic curve showing solar sales-volume against unit price: the higher the sales of solar go globally, the lower the price goes. At the time, of course, with minimal sales volume, the unit price was high. A line on the graph hit the price–volume curve at a point in the future where global sales had grown considerably, and prices had dropped to a particular level. Shell, he said, had decided it would jump into the solar market at this particular point on the curve, and buy up everything it could. Of course, in the meanwhile, it would do nothing to help the solar market speed its descent of the curve. It would carry on concentrating on its core business until somebody else figured out a way of bringing the sustainable future within reach of the unsustainable present.

This is the same brand of cynicism that allowed BP and Shell to back the carbon club. Several times, at meetings with the companies since 1991, I had urged senior executives to quit the Global Climate Coalition. They prided themselves on their rigour in technical arguments, I would say, so how could they go on supporting – indeed funding – a crass disinformation campaign run by crude bullies? Weren't they just storing up problems for the future?

Shell in particular prided itself on being an oil company able to look constructively into the future, and coordinate the global projection of a clean, coherent, image. According to the founder of Shell's scenario planning process, the company needed to guard against 'the parochialism of the internally constructed version of reality'. As a former Shell head of strategic planning put it, 'it is extremely difficult for managers to break out of their world-view while operating within it. When they are committed to a certain way of framing an issue, it is difficult to see solutions that lie outside this framework.' Scenario planning had helped build Shell a formidable reputation as a forward-looking corporation by 1995.

Of course, all that was in the pre-*Brent Spar* world.

The *Brent Spar* was the first oil-storage platform in the North Sea to need decommissioning. It was an installation in the British sector, co-owned by Shell and Exxon but operated by Shell UK, which had been installed in 1976 as an offshore repository for oil from the giant Brent field, and abandoned in 1991. At the time the *Spar* began its life, Shell, like all the other oil companies in the North Sea, was promising fishermen and governments that they would leave the seabed exactly as they had found it.

For years, Shell UK had run a successful PR campaign in the United Kingdom called the 'Better Britain' campaign. This consisted of an annual set of awards for conservation schemes. Shell was effectively exhorting the British population to act with environmental responsibility, including when dealing with garbage. So what did the company plan to do, in May 1995, with the 65,000 tonnes of steel and concrete, plus over 100 tonnes of toxic waste – including oily sludge, arsenic, cadmium, PCBs (polychlorinated biphenyls) and lead – that was the disused *Brent Spar* platform? It planned to dump the lot in the sea.

In terms of precedent, the company was clearly running a dangerous gauntlet from day one. The US Government had for a long time insisted that all oil platforms in its coastal waters be towed to land and dismantled. Since 1987, 914 structures had been decommissioned in the Gulf of Mexico in this way. But Shell UK now argued that the *Brent Spar* was one of an undisclosed number of the four hundred North Sea structures for which an exception should be made. The best thing to do was to tow the rig into the Atlantic and dump it in deep water, they argued, toxic waste and all. The British Government agreed, and issued the necessary licences.

On 30 April, Greenpeace climbers boarded the *Spar*, announcing that they would stay there until Shell changed its mind. The activists maintained an arduous vigil aboard the platform for over three weeks. It was far from evident in these early days that this was an action that would end up as a landmark in environmental politics. There was little press coverage outside the UK, Germany, and the Netherlands at first. But by the time that Shell UK officials, accompanied by police and sheriff's officers, were put aboard the *Brent Spar* on 24 May, things had begun to change. The eighteen activists and journalists by then aboard the rusting platform were ordered to leave, without arrests. But if Shell thought that was to be the end of the problem, they were spectacularly wrong. By this time, Shell and the British Government had been publicly condemned by the European Parliament, the European commissioner for the environment, and the Belgian, Danish and Icelandic governments. More problematically for the company, regular press coverage had started in many European countries.

Secretary of State for the Environment John Gummer now had to defend his government's cosy collusion with Shell at the North Sea Ministers' Conference, between 7 and 9 June, in Denmark. This was an event to which Greenpeace's occupation of the platform had of course been designed to draw attention. Gummer had a miserable time. The UK was once again under fire as the 'the dirty old man of Europe', as environment ministers from other European countries played strongly to green constituencies back home at his expense. To add insult to injury, he was faced with evidence that Shell had been economical with the facts in its dealings with the British Government. A Shell report leaked during the conference revealed that top engineers Smit Engineering had advised the company that the *Brent Spar* could be towed back to land and safely decommissioned just as cheaply as it could be dumped at sea.

On 11 June, with the ink barely dry on a North Sea Ministers' Declaration roundly critical of sinking oil platforms at sea, Shell incomprehensibly began to tow the *Brent Spar* off towards its deep-sea grave. Greenpeace tailed the *Spar* in two ships, one of its own and one chartered, and highly televisual images began appearing on news bulletins across Europe. By now public opposition to Shell's plans was growing apace. Across Europe, motorists were by now avoiding Shell filling stations in their thousands. On 13 June, the top-circulation German paper *Das Bild* called for a boycott of Shell on its front page.

On 14 June an astonishing thing happened: public dispute erupted within Shell. The Shell Netherlands president said in a TV interview that he would be asking Shell UK and the British Government to reconsider their decision to dump the *Spar*. Shell Austria's managing director said on national radio that he didn't agree with the dumping. It turned out, incredibly, that Shell UK had not sought the opinion of its European sister companies when it had elected to dump the platform. 'We knew that this would play disastrously in Germany,' one Shell Germany executive told the press. He would have told his colleagues as much, except that the first he had known of the dumping was when he heard about it on television.

Greenpeace activists aboard the Spar now made a sampling error when endeavouring to find out how much toxic material was aboard the structure. The campaigners in charge, by then exhausted, did not have normal quality-control measures in place, and on 15 June a horrible overestimate was released to the press. This provided Shell and others with a wonderful weapon to use in the later post-mortems on the *Brent Spar* affair. To this day, many people believe that Greenpeace sampled incorrectly right at the outset, and based the entire campaign on a hopelessly inaccurate analysis. In fact, until the erroneous estimate was released eight days from the eventual campaign victory, Greenpeace's figures for the toxic material aboard the *Spar* had been based on Shell's published estimate of 50 tonnes.

On the same day, at the G7 summit in Canada, Germany's Chancellor Kohl publicly urged UK Prime Minister John Major to reconsider the planned dumping. Major refused to budge. The German finance minister spoke in anger to reporters. 'What's the use of international conferences on the environment,' he railed, 'when things like this take place which are incomprehensible to everyone?'

The *Brent Spar* story was now very much world news, and the cameras were still running.

Public anger was huge. In Germany, a violent fringe went so far as to fire-bomb a Shell filling station, and shots were fired at another. But Shell had succeeded in awakening not just the wrath of a lunatic few: they had sparked off a genuine grass-roots rebellion. The consumer boycott of Shell's petrol had cut Shell Germany's takings by 20 to 30 per cent. Opinion polls showed that four in five Germans favoured the boycott. Europe-wide, the company was losing up to $10 million a day.

On 17 June Shell was scheduled to hold its annual 'Better Britain' environmental awards ceremony in London. It was postponed. The pressure and humiliation had become unbearable. The company now faced the real prospect of lasting and unpredictable harm to its image. It was not so much the lost revenue: a company which had global sales in 1994 of well over $100 billion could afford to take a few losses in the tens of millions range. It was the loss of image that was the real problem.

Late on the afternoon of 20 June, barely a day after John Major had told the UK parliament that dismantling the platform on land was 'an incredible proposition', Shell announced that it had dropped its plans to scuttle the *Brent Spar* at sea. Out in the Atlantic, the Greenpeace ships trailing the Shell flotilla watched the rusting hulk stop, and begin to turn.

In those moments, recorded on film by Greenpeace Communications amid the cheers of the watchers, a rainbow appeared in the sky.

Business-school textbook chapters will be written on the lessons of the *Brent Spar* episode for years to come. Let me offer just two. The first and biggest lesson is that industry in general, and the oil industry in particular, encountered a clear turning point in environmental politics on 20 June 1995. Although the risk to life in the incident was nothing compared to Chernobyl, or Bhopal, this was clearly as seminal a political episode. The world's heavyweight papers seemed to be in broad accord on this. 'People no longer accept that the men from the ministry, let alone the multinational, know best,' the *Financial Times* concluded in an editorial which captured the general vein.

The second lesson concerns the apparent built-in propensity of the oil industry to underestimate its environmental liabilities. The *Brent Spar* episode means that dozens of similar installations in the North Sea will have to be decommissioned on land, at a cost of around $1.5 billion over the next ten years. If smaller structures – ordinary drilling rigs, rather than oil-storage platforms – have to be decommissioned in the same way, the total cost will rise to $7.5 billion or more.

The critical question that oil companies must ask themselves is whether or not this propensity extends to other areas of the industry's operations. To the ageing fleet of creaking oil tankers? To the tens of thousands of kilometres of leaking oil pipelines? Even to the enhanced greenhouse threat?

JUNE 1995, KYKUIT, NEW YORK STATE

I am sitting at a bed-sized desk in a barn-sized room on the top floor of a cavernous granite mansion. The wind is rushing off the woods down along the Hudson River. It blasts at the green fleeces across the four-hundred-acre grounds and moans through the hardwood window frames. Otherwise the house is deep in quiet, and as I write these words a shiver passes down my spine. I am looking at a photo on the desk. It shows an old man, looking like many an old man, with sunken cheeks and thinning grey hair neatly parted above a high brow. He could be anybody's grandfather. Perhaps the eyes are cold, the thin lips cruel, but these are clichés and probably also imagination. I did expect to see more character in the features, though, even if he was 91 when the picture was taken. For he is John D. Rockefeller, the founder of Standard Oil and hence of Exxon. And it is his house I am in.

I have come to the mansion at Kykuit not to enjoy the Rockefeller family's rose gardens, or the views across the woods to Manhattan, but to plot with a group of Americans how best to put climate change on the agenda in their country. We come from the churches, the cities, the foundations, the universities, the business community and, of course, the environmental groups. We have been installed here for two days by the Rockefeller Brothers Foundation, our hosts. We have bedrooms in the mansion itself, but we are working in the conference room down in the old coach house, below the museum where yesterday I saw my first Model T Ford, the car that propelled Rockefeller's oil juggernaut into the fast lane.

The richness of the irony has me in thrall. I am in the home of the father of the oil era, and I am conspiring to plot the fall of oil.

What would the old man have made of our gathering? Rockefeller was by all accounts a deeply religious man, strong on ethics. Yet that made him a man of contradictions, for in becoming one of the richest men the world has ever seen, he created a ruthless global monopoly, and ran it using – among other tactics – a network of secret agents and widespread bribery of officialdom. On the other hand, so the guidebook on my desk tells me, charity was an essential part of

Rockefeller's life. By the time he died, aged ninety-seven, Rockefeller had given away over half his fortune to philanthropic programmes.

I wonder what Rockefeller the Christian and philanthropist would have thought of the suggestion that the product he was delivering to the world would end up as a threat to much of Creation and to the future of civilization? To the north of here, huge areas of Canada's boreal forests are ablaze. To the south, Midwest cornfields normally basking in sunshine are deep under water. Here in New York State, the arid summer is drying up rivers and cutting hydroelectric power output drastically. How would Rockefeller have reacted to the prospect that these problems could, in major part, be a result of the burning of his product, and other fossil fuels? He was a man who evidently valued knowledge as highly as ethics. In 1913 he set up the Rockefeller Foundation, one of the world's largest foundations, 'to promote the well-being and to advance the civilization of the peoples of the United States . . . and of foreign lands in the acquisition of knowledge, in the prevention and relief of human suffering, and in the promotion of any and all the elements of human progress.'

Worthy goals indeed, and the pollsters are here at Kykuit, as background to our deliberations, to tell us about the state of environmental ethics and knowledge in America today. They have told us that 78 per cent of Americans agree with the view that because God created the natural world it is wrong to abuse it. Similarly, 93 per cent of Americans feel that working hard to prevent environmental damage for the future is part of being a good parent, while 62 per cent even think that protection of the environment should be ranked ahead of economic growth, in terms of social priorities. These ethical values are little different, in essence, from those revealed by polls in Europe. Following Brent Spar *there is much talk about Europeans being more environmentally conscious than Americans. But it seems to me that things are little different on the two sides of the Atlantic: the* Brent Spar *episode merely awoke a force in Europe which in the USA has yet to encounter an issue capable of jolting it out of its slumber.*

What about perceptions of global warming, and knowledge of the greenhouse effect? Here things are certainly different. In Germany, 73 per cent evidently consider global warming to be a very serious threat; in the USA, only 43 per cent think likewise. Ranking perceptions of the seriousness of global environmental issues, the pollsters tell us that Americans place global warming last behind toxic waste, loss of rainforests, water pollution, air quality, overconsumption of resources and rapid population growth, in that order. On the other hand, fed the statement that 'every time we use coal or oil or gas we contribute to the greenhouse effect', 30 per cent of Americans say they think it's not true.

Confusion like that can usually be put down to the disinformation peddled by the oil companies, not least of them the company that Rockefeller founded. Had he gone to the climate talks, and seen men like John Schlaes and Don Pearlman giving stalling briefs to OPEC negotiators and finger-wagging the US delegation, or Brian Flannery and Fred Singer deploying their smokescreens in the climate-science debate, would he have been surprised?

Sitting on my temporary desk in Rockefeller's home is a report. It is by a member of the Lloyd's of London delegation to the Berlin Climate Summit, and even now it is circulating, sitting on similar oversized desks, at high levels in the European and US insurance and banking industries. It contains some interesting conclusions about the implications for insurers of the Berlin Summit. 'The obligation of managing risk, both for the benefit of their customers and for their capital providers, means that not to monitor scientific developments and to ignore the initiatives of the IPCC and most fundamentally not to have their voice heard and their interests not represented, except as purportedly expressed through the International Chamber of Commerce and the Global Climate Coalition, is a questionable exercise of their responsibilities.'

Strong stuff, and bang on target. Looking into the future, the report further concludes that it is inescapable that the insurance industry is going to have to take some initiatives by itself, perhaps in concert with the banking industry. And when it does so, the markets will be sure to listen.

The race against time is on, and we should have started running by now: racing to wind down the oil era, racing to crank up the solar revolution, racing to mobilize the capital markets as engines for survival instead of suicide.

I'm cautiously optimistic, but I'm also scared. And the more we mill around on the starting line, the more scared I get.

Now, as I look down at my computer screen for the last time, I hear only the sound of the heaving woods out by the Hudson and the deep silence in the house of the father of oil. I feel a breath of them on my cheek, and I sense the ghost of Rockefeller looking over my shoulder.

And I have a fancy that he is scared too.

8

A Discernible Human Influence

May 1995–June 1996

AUGUST 1995, STAVANGER, NORWAY

BP's head of health, safety and environment, Klaus Kohlhaus, must have recounted his company's toxicological vital statistics proudly to audiences at oil-industry conferences many times. The performance between 1990 and 1994 was one of constant improvement, he told over a hundred industry people at Norway's annual oilfest. Refinery discharges to water: down 55 per cent. Refinery hydrocarbon emissions to air: down 18 per cent. BP Chemicals discharges to water: down 60 per cent. BP Chemicals hydrocarbon emissions to air: down 41 per cent.

Then Kohlhaus came to climate change. Better science was needed, it seemed, before anyone could think about emissions cuts.

I knew Klaus Kohlhaus to be a decent man. We maintained cordial relations, despite our public clashes. I wondered how he could say it without sounding sheepish.

I took my turn on the platform. It was as though BP had segmented its thinking on the environment into different lobes of the corporate brain, I volunteered. I conceded that it would be churlish not to give the company credit on its real environmental achievements. But these achievements applied in every area *except* carbon dioxide. It really was as if the company had a huge blind spot when it came to the greenhouse issue. The science was no more uncertain than that behind some of the toxics issues. Yet carbon dioxide was viewed as a non-problem.

In the subsequent panel discussion, sitting at a table on the platform, Kohlhaus and I spent most of our ritual debate on the question of climate science. BP, he emphasized, was not persuaded that there was a serious problem.

The time came for questions from the audience.

After some fairly predictable arrows had come my way, I noticed an extremely elegant man near the front put his hand up. It was an insurer, Carlos Joly of Storebrand, Norway's biggest insurance company. He had been listening to the arguments, he said, and had come to the conclusion that he agreed with the man from Greenpeace. He wanted to explain why.

'I am an investor. I work for the biggest investor on the Oslo Stock Exchange.' Joly very much had the audience's attention at this point, I observed. 'I am used to investing in the face of risk. I never have 100 per cent certainty that my investment is a good one. So I know something about odds, and I have to say that – sadly – if I had odds on an investment being a winner as strong as seem to favour global warming being a problem, I would not hesitate to make that investment.'

Klaus Kohlhaus had debated with me throughout our long earlier exchanges on the basis that the climate science was still unproven. But now, as soon as he was faced with this approach, he switched tack. He did not challenge Joly's risk point, but strung together an answer to the effect that the world had been run on oil for a long time, was still run on oil, and was likely to be run on oil for a long time to come. 'Are you telling me something different?'

The closing question angered Carlos Joly. 'I *am* telling you something different.' This said slowly, with voice raised.

Joly continued more calmly, picking his words carefully. 'I am telling you that if public concern over global warming increases anything like the concern over dumping of oil platforms has increased, then BP is not going to be as interesting an investment prospect in the future as it is today.'

I risked a look along the table. Kohlhaus was scarlet, whether with anger or embarrassment I could not tell. How often had he been spoken to like that, I wondered, in front of an audience of his peers?

But to his great credit, he came up to me immediately afterwards. We had not had time to say hello before the session, and now he shook my hand.

'Jeremy,' he said, 'you are doing well with this insurance business. Very well. I will have to tell my board about what has just happened, you know. We can't afford to ignore something like that.'

The next phase of climate talks – those due to take place between the recent first annual Conference of Parties in Berlin, and the goal for adoption of emissions targets, the Third Conference of Parties in 1997 – was to become known as the Ad-hoc Group on the Berlin Mandate. Hence, AGBM would be the next set of initials to add to the growing list of climate acronyms. The first such set of negotiations, AGBM1, took place in Geneva in the same week I was engaged with the oil industry in Norway.

Unlike my time in Stavanger, my colleagues in Geneva endured a week of boredom and frustration. By the end of the session, the governments had not even agreed on the composition of the bureau for the post-Berlin talks – the important group of governments informally advising the Chairman. The Berlin Mandate called for an 'analysis and assessment' phase, during which developed countries would decide on the policies and measures they would use to cut greenhouse-gas emissions. It also called for talks to set 'quantified emission limitation and reduction objectives' (another acronym to grasp: QELROs), with the aim of agreeing a deal by the Third Conference of Parties – to be known universally as COP3. The EU, and many developing countries, stressed the fact that COP3 was only two years away. They wanted QELRO negotiations on a protocol or other binding agreement to strengthen the convention in parallel with the analysis and assessment phase. The USA did not. That was putting the cart before the horse, they insisted.

NOVEMBER 1995, KNOSSOS, CRETE

In the ruins of Knossos, the centuries have reduced the royal box in the palace theatre to nothing more than a knee-high platform of shoe-worn stone blocks, but the view down the sunken pine-lined avenue to the west is of the oldest road in Europe, and if you squint in the evening sun you can imagine it as it was, feeding into the nerve centre of the Minoan civilization.

As the sun slipped behind the olive groves, I sat resting, where the Minoan kings once sat, reading my Anne Rice novel. I needed escapism that day. A rare anger had lodged in me from the moment I heard the news that the Nigerian government had executed Ken Saro-Wiwa. While the Commonwealth's heads of state were sleeping at their summit in Auckland, secure in their belief that 'quiet diplomacy' would work with General Abacha, the latter was busy consummating his process of juridical murder. The thorn in Shell's side, the pillar of the Ogoni people's struggle against oily oppression, had swung his reluctant way to a martyr's death.

A mighty public-relations war had built up around the world during the months of Saro-Wiwa's incarceration, and his sham trial. In the UK, for example, Greenpeace, Friends of the Earth and the Body Shop paid for full-page newspaper adverts decrying the despoliation of Ogoniland by Shell and the Nigerian Government, and exhorting Shell to intervene with the General on behalf of Saro-Wiwa. Shell, under attack once more in the editorial columns, had replied with a PR blitz of their own. They, like the Commonwealth heads of state, stressed the need for quiet diplomacy. I wondered what they meant by that. Shell produces half the oil in Nigeria. Could it have been that behind the scenes they quietly warned the good general and his gold-braided henchmen that some of Nigeria's $12 billion a year oil income might have been at risk? I did not think so.

En route to Amsterdam the next day, front-page headlines would tell me that Shell was to sign a $4 billion deal with the Nigerian Government for a huge new gas plant.

Knossos. A place to think about civilizations. The rulers of the Minoan civilization walked through more than a thousand rooms and a labyrinth of corridors in this palace. Two thousand years before Christ they brought water from Mount Zeus, 19 kilometres distant, in precisely aligned terracotta pipes, feeding among other things Europe's first flushing toilet. As many as 80,000 inhabited the surrounding city, and at its height the Minoan civilization projected naval power throughout the then known world. Elaborate frescos of underwater scenes and slim-waisted dancers suggest a love for nature and a zest for life, and in the main square near where I sat that day young acrobats, male and female, had played dangerous games with bulls while the royals looked on.

But civilizations die. In 1450 BC the volcano Santorini, a hundred kilometres away, blew its top in one of the biggest eruptions in recorded history. A tidal wave swept over Knossos, which was then right on the Cretan coast, and the death-blow to the Minoans was compounded by a devastating fire, scorch marks from which can be seen on alabaster and gypsum blocks throughout the ruins.

The Minoan civilization, into its twenty-fifth generation or thereabouts, had no warning, no chance of risk abatement or avoidance. The fifth generation of the oil civilization has abundant warning. An approaching apocalypse is building in increments, clearly visible for those with eyes to see. I came to Crete from Athens, where the children of the oil civilization are slowly killing themselves and their fellows in thousands, notwithstanding global warming. In the cavernous and cramped streets, to breathe is like dragging on a cigarette laced with diesel particulates and benzene. And Stelios Psomas, my ever-optimistic Greek colleague, told me that we had had a lucky week. A wind was blowing most of the filth out into the Aegean.

Stelios and I thought we had made some progress in our effort to keep the waves and fires at bay. At a conference on renewable energy the European energy commissioner, Christos Papoutsis, stated that the world desperately needed rapidly growing renewables markets. Crete could be a model for the way forward, he said. In a room at the hotel Grande Bretagne, we congratulated Papoutsis on his vision and presented him with a proposal for beginning the process of building his cradle. Two nights before, I had spoken with Bob Kelly, who had set up another subsidiary of Enron, Amoco-Enron Solar Power Development Corporation, of which he was chief executive. I told Kelly about a progressive new renewables law in Greece, allowing a guaranteed payback of 10 cents per kilowatt-hour for electricity generated by renewables schemes. Crete had an electricity crisis: a projected growth in demand of 8 per cent per year. The monopoly state utility known as the Public Power Corporation (PPC), replete as it was with Guardians Of All That Is Old And Traditional, had in its wisdom decided to solve this problem by building a 150-megawatt oil-fired power station, progressive renewables law or no progressive renewables law. An oil-fired power plant in the land with the highest insolation in southern Europe! What could I tell the Greek government, and the Greek European energy commissioner, about Enron's interest in this state of affairs?

Kelly told me that if he could build one of his photovoltaic power stations on Crete, he would also build a PV production plant in Greece. He would come to Greece to tell them that himself, and with details, if there was any interest.

Interest, predictably, there was. Stelios and I met the environment minister. He would love to talk to Enron, he said, and he knew his Cabinet colleagues would too. Who wouldn't be interested in such an investment coming to Greece? And jobs as well. Besides, the local people on Crete were opposed to the idea of a dirty oil-fired plant, the minister understood.

Stelios knew this very well. When he had debated with PPC officials who had gone to Crete to outline the scheme, crowds of locals had greeted the utility men with black flags and chants, and Stelios with flowers and smiles. That was a mere foretaste of what we could do, Stelios assured me, if PPC went ahead with their plan.

Hearing all this, the European energy commissioner began at once to plan ahead in principle. Europe's first large solar PV power plant could be used as a platform for vertical proliferation of solar power within Greece, he said, and horizontal proliferation within Europe. Listening, I began to feel optimistic that we might have a real chance, for the first time anywhere, to prevent the construction of a planned fossil-fuel power plant, and replace it with a solar scheme.

Thus had I come to my one-day break on Crete. A Sunday sitting in the ruins of a lost civilization, in a region which had been the cradle of so many other civilizations in the 3,000 years since.

Perhaps again now? The solar civilization?

Through the summer, a group of European and Japanese insurance companies had been meeting at the behest of UNEP to try to negotiate a joint statement expressing their concern about the global state of matters environmental. General Accident and National Provident

Institution of the UK, Gerling-Konzern of Germany, Swiss Re, Storebrand of Norway, and Sumitomo of Japan, finished the task in September in Cologne. In November they released their Statement of Environmental Commitment by the Insurance Industry in Geneva. By then six other companies had joined them, and they appealed to the industry at large to sign up.

The use of the word 'commitment' in the title was significant. The principals intended the statement to have meaning, in that once a company signed it would feel itself to be under real pressure to move forward with proactive measures. An equivalent and more general Statement by Banks on the Environment and Sustainable Development, also brokered by UNEP and signed back in 1992, seemed with the benefit of hindsight to have provided too little incentive for most of its signatories to act.

The insurers' statement contained some interesting passages. 'We are committed to work together to address key issues such as pollution reduction, the efficient use of resources, and climate change,' the preamble stated. And the specifics included the following: 'we will seek to include environmental considerations in our asset management'.

At last the investment-in-carbon issue was right out in the open, acknowledged by at least some insurance companies.

NOVEMBER 1995, MADRID

During 1995, climate scientists were becoming progressively more willing to say that a human global-warming signal was appearing. In April, a researcher from AT&T's Bell Laboratories published a statistical treatment of the twentieth-century warming record that showed a correlation between warming and a decreasing amplitude of the annual temperature cycle – the temperature difference between winter and summer. Had changes in incoming solar radiation been solely or largely responsible for observed warming, as greenhouse sceptics tend to argue, the reverse should have been the case. AT&T researcher David Thomson told *New Scientist* magazine that 'you almost have to invoke magic if it is not carbon dioxide. It's the only logical explanation.'

In May, scientists at the US National Climatic Data Center reported that they had found a pattern suggesting that the US climate had moved towards a greenhouse regime in the last 15 years. They compiled data on climatic extremes expected to arise under greenhouse warming, including higher-than-normal daily minimum temperatures, extreme or severe droughts in warm months, and much higher-than-average precipitation in cool months. Using these data, they calculated a 'greenhouse climate response index' in order to quantify any persistent trends. The index has been consistently high since the late 1970s, in marked contrast to the pattern in earlier decades. They concluded that the probability of this being natural fluctuation in a stable climate was only about 5 per cent.

In June came two other high-powered studies similarly dismissive of natural climatic variability as an explanation for the hot years of the 1980s and 1990s, this time in the global record. A team from Germany's leading climate lab, the Max Planck Institute for Meteorology, concluded that there was only one chance in forty that natural variability could explain the warming seen over the last thirty years. A second team, at the USA's Lawrence Livermore Laboratory, factored the cooling influence of sulphate aerosols (largely derived from coal burning) into the temperature record, leaving a clear greenhouse fingerprint since about 1950, when the growth of carbon dioxide emissions took off. The correspondence between model and actual temperatures was good geographically, seasonally and vertically in the atmosphere. The team leader was Dr Ben Santer, who was destined to play an unwilling role centre stage in the months ahead.

In August, the results of the most sophisticated climate-model experiments yet performed were reported in *Nature* magazine. Researchers at the UK's Meteorological Office simulated past climate variations using a coupled model (one linking both the atmospheric and oceanic components of the climate system) which also factored in the sulphate aerosol effect. This breakthrough significantly boosted confidence in the accuracy of the models used for forecasting climate change in the future. Leading IPCC climatologist Tom Wigley wrote in a commentary that 'these results mark a turning point . . . in our ability to understand past changes and predict the future.'

And at the same time, the drip of worrying evidence that the feedbacks may be preparing to kick in, stoking up an amplifying effect for future warming, continued. During July, my long-standing fears about methane hydrates resurfaced when I read that Geological Survey of Canada scientists taking cores in the MacKenzie river delta had found hydrates at much shallower depths than had been thought possible. Their concern that this could boost the greenhouse effect featured prominently in *New Scientist*.

During August, a team of researchers at the US National Oceanographic and Atmospheric Administration reported that they had found isotopic evidence for a large sink for carbon dioxide inland in the northern hemisphere during 1992 and 1993. We should worry about this, they concluded. Carbon storage in this land sink would probably be more vulnerable to greenhouse warming than carbon storage in the oceans.

Within a week, as though to abate hopes that even the oceanic sink would hold up well, came evidence in *Nature* magazine that the oceans may be losing fixed nitrogen. Fixed nitrogen, as the nutrient nitrate, is a major control on the abundance of phytoplankton. Denitrification – the degradation of nitrate by bacteria in areas of oxygen-deficient surface waters in the oceans – converts fixed nitrogen to free nitrogen or nitrous oxide, of no use to most plants. Studying nitrogen balances, Canadian and American oceanographers found evidence that warmer oceans will lose nitrate, so depleting phytoplankton, so boosting atmospheric carbon dioxide.

The IPCC had been busy throughout 1994 and 1995 preparing its all-important Second Assessment Report. If scientific concern about global warming was abating, the steam would be taken out of the negotiations. If, as the contents of the premier scientific journals suggested, reasons for concern were in fact compelling, then governments might just be persuaded to focus their sights on emissions. The scientific working group's report was nearing completion in early November. And as always seems to happen with multi-author reports, somebody leaked it to the *New York Times*. The *Times* ran a front-page story. The article focused on one chapter of the long report, on the increasing correspondence between predicted patterns of climate change and those actually being observed.

The headline was 'EXPERTS CONFIRM HUMAN ROLE ON GLOBAL WARMING'.

The leading IPCC scientists gathered in Madrid late in November to finalize their part of the IPCC's Second Assessment Report. I knew that the carbon club would have to try to stop the draft report becoming an official one.

The 150 or so IPCC scientists scanning the room on the first morning of the meeting had some new faces to look at, but they were not new to me. Fred Singer was a scientist of sorts, but he had attended neither of the two previous IPCC scientific plenaries. Don Pearlman did not even pretend to have scientific credentials. He had attended the 1992 plenary in China, where he failed to have any impact. In Madrid, however, he had heavier guns to hand than he had had in Guangzhou. Two new Global Climate Coalition lieutenants, Drs Olaguer and Gardner, both seemed to know one end of a climate model from the other, and if they needed advice on where to aim their torpedoes, Leonard Bernstein, Mobil's head of global warming damage limitation, was there to give it. Most worrying of all, several Saudi and Kuwaiti delegates from the climate negotiations had turned up. Not only were these gentlemen not scientists, they were not even diplomats: they were oil ministry officials. One of them was Mohamed Al-Sabban, long-time leader of the Saudi delegation, a man with years of experience of how to frustrate progress and slow the pace of climate negotiations.

The man who would have the terribly difficult job of chairing this working-group plenary was John Houghton. He had handled earlier plenaries in Berkshire in 1990 and China in 1992 with aplomb, but marshalling this particular 40-page draft into an agreed form of words in just three days was probably going to be the biggest challenge of his life.

After the reports by lead authors, and questions to them, the meeting was called to order for its first work on the text itself. Immediately, the golden first paragraph of the report came under attack. It was unacceptable to say that 'emerging evidence points towards a detectable human influence on climate', said Al-Sabban. The chapter of the full report had used the word 'preliminary' in this context.

Kuwait supported this. So did Kenya, Pearlman and the Global Climate Coalition.

Al-Sabban's tactic was crude but effective. Tom Wigley, a senior scientist at the US National Center for Atmospheric Research and a lead author, was nonplussed at the Saudi approach. He, like many of the scientists there, had not seen this kind of thing before. 'Preliminary', in the context to which Al-Sabban referred, was manifestly not intended to imply that evidence of human influence was uncertain. Rather, it had been used by the authors in the sense of initial, but none the less clear and unambiguous. 'We did not realize how this word could be misinterpreted,' he would later tell *Nature* magazine.

John Houghton decided to move swiftly on to less contentious ground, and return to the key conclusion later. But with a subtle little lawyer's ruse, it looked as if the die had been cast for the days – and months – ahead.

On the first evening, the Spanish Government had organized a reception. As I chatted to the head of the US delegation, Bob Watson, about the problems facing the Chairman, Fred Singer sidled up to me.

'Jeremy Leggett, as I live and breathe.'

'Hello, Fred. I haven't seen you since you lied to the *Washington Times* about me.'

Bob Watson cleared his throat and moved off before I could explain about Singer's treatment of my worst-case opinion survey.

'How are you doing?' Singer continued without bothering himself with a denial. 'Written any books lately?'

'And you? Unaffected by the recent contents of *Nature* and *Science*?'

The old man was in jocular mood. 'No, I am more than ever sure that I am right.'

'What I regret,' I said, 'is that you won't live to see how wrong you are.'

We fenced on in this vein. Then he came back to my opinion survey, or at least that was what I presumed he was referring to.

'I hope I didn't hurt you too much.'

Not being of the Oscar Wilde school of wit and repartee, this was time for me to either blow a gasket or walk away.

I turned away, with a parting shot that fell well short of the minimum entry requirement for the *Oxford Book of Insults*.

For the next two days the battle over detection, and much else in the draft report, dragged on. There is a constant danger, in recounting the history of these events, of repetition. Repetition may deter readers, but in the long history of the climate negotiations it has never deterred the carbon club and its government proxies.

Suffice it to recount the following sequence of drafting changes. The original draft had a section heading on detection entitled 'The changes point towards a human influence on climate'. The draft the next day read 'The balance of evidence points towards a human influence on climate'. At one point, Bob Watson proposed 'The balance of evidence suggests an appreciable human influence on global climate'. By the time the meeting finished, the same title read 'The balance of evidence suggests a discernible human influence on global climate'. As for the header right at the top of the executive summary, 'Significant new findings since IPCC 1990', that had disappeared completely.

This watering-down was achieved entirely by the carbon club, the Saudis and the Kuwaitis, who knew the scientists would have to exercise compromise while at the same time racing against the clock. True, they did not achieve their complete objective, or anything like it. At one point Al-Sabban, the oil ministry official, proposed to a hundred-plus of the world's best climate scientists the following: 'Preliminary evidence which is subject to large uncertainty points towards a human influence'. He did not get his way with that. But this was not the point. The point was to wear the scientists down with diplomatic gutter tactics, if possible to derail the meeting procedurally, but certainly to reduce the impact of the product.

This, to a significant extent, they succeeded in doing. The final product was not 40 pages as it started, but a mere 11, with the rest as a technical appendix. Vital sections of text, and crucial diagrams of the kind that had been so important in the first IPCC scientific assessment, had been denied the official blessing of an IPCC plenary.

The session on the final day extended until past midnight. Much earlier, with Houghton struggling impossibly against the clock, NGOs had been asked to stop making interventions and leave the floor to governments. By this time, many scientists – not having the stamina of diplomats – had departed. As the room thinned, Don Pearlman became progressively more brazen as he ferried written notes to Al-Sabban. The Saudi would read them, and up would go his flag for another filibustering move.

The US delegation, desperate to reinject some of the excised meat from the report, suggested that a short new section title be included. It would read 'Only reductions in emissions will achieve greenhouse gas stabilization'.

'We do not have a quorum,' said Al-Sabban instantly. 'We cannot agree this as a minority.' And Chairman Houghton, visibly exhausted and stressed, conceded the point.

The draft of the report entering the last day had a section entitled 'There are still many uncertainties and possible surprises'. Al-Sabban now asked that this be amended to a simpler form: 'There are still many uncertainties'.

He got it.

Al-Sabban would later tell *Nature* that the report 'is now balanced and cautious enough not to mislead policymakers'.

In contrast, a senior researcher from the US National Center for Atmospheric Research, Dr Kevin Trenberth, asked towards the end of the Madrid meeting if he could have his name removed from the report as a lead author.

When pressed on his actions by *Nature* magazine, Al-Sabban had this to say: 'Saudi Arabia's oil income amounts to 96 per cent of our total exports. Until there is clearer evidence of human involvement in climate change, we will not agree to what amounts to a tax on oil.'

Through it all, the lead author of the detection section, Ben Santer of the Lawrence Livermore Laboratory, had been in the thick of the discussions and enforced redrafting. It was he who had to deal with many of the brickbats from the carbon club and their proxies. This he did in a dignified and patient manner. At the end of the meeting, Santer was duly told to go back to the original full report and make appropriate changes to match the changed Policymakers Summary.

The carbon club would be able to turn even this simple instruction into a grenade, as we shall see.

DECEMBER 1995, CITY OF LONDON

The IPCC's vital Second Assessment Report was completed on 15 December in Rome, at a meeting I did not attend. The reports showed that, with all governments in attendance and many alert to what had happened in Madrid, the carbon club was much less effective than it had been the previous month. The general mood among those to whom I spoke after Rome, both from governments and NGOs, was that the overall IPCC report as completed would have a good chance of sending the signal that was needed to stiffen resolve in the climate talks proper.

In brief, the scientific section confirmed a mid-range estimate for global warming of around 2°C by 2100, with a low-to-high estimate range of 1 to 3.5°C for the emissions scenarios considered. These figures factored in the aerosol dampening effect, which the first IPCC scientific assessment hadn't. The sea-level rise mid-estimate was 50 centimetres by 2100, with a 15 to 95 centimetre range. All this, the impacts section made clear, ought to be of great concern. 'Human health, terrestrial and aquatic ecological systems, and socio-economic systems (e.g., agriculture, forestry, fisheries and water resources) are all vital to human development and wellbeing and are all sensitive to changes in climate,' the report concluded, and in all these sectors 'potentially serious changes have been identified.' Some of the conclusions on impacts made very depressing reading indeed. 'Entire forest types may disappear . . . large amounts of carbon could be released into the atmosphere . . . desertification is more likely to become irreversible . . . between one-third and one-half of existing mountain glacier mass could disappear over the next hundred years,' and so on.

'Climate has changed over the past century,' the report stressed. Notably, warming has been greatest over the mid-latitude continents in winter and spring, and precipitation has increased over land in high latitudes of the northern hemisphere. Moreover, 'climate is expected to continue to change in the future,' including an increase in the occurrence of extremely hot days, more severe droughts and/or floods, plus, according to several of the climate models, an increase in precipitation intensity.

The remaining uncertainties, the report made clear, did not necessarily give less cause for concern. Far from it. 'When rapidly forced, non-linear systems [like the climate] are especially subject to unexpected behaviour.' Examples include rapid circulation changes in the North Atlantic and feedbacks associated with terrestrial ecosystem changes. And the report went further than its 1990 predecessor in spelling out this danger. 'Future unexpected, large and rapid climate system changes (as have occurred in the past) are by their nature difficult to predict. This implies that future climate changes may also involve "surprises".'

On the possible policy response, the 1995 report went much further than the 1990 one. 'Significant reductions in net greenhouse-gas emissions are technically possible and can be economically feasible,' it concluded, 'using an extensive array of technologies.' In 1990, the IPCC had been steered away from anything like this by the Bush Administration. Moreover, the report concluded, 'earlier mitigation may increase flexibility in moving toward stabilization of atmospheric concentrations,' and 'significant "no regrets" opportunities are available in most countries.'

A major difference between the first and second IPCC reports was that the second had an entire chapter about the implications for the financial sector. There was little that could be said by the scientists that was of specific help to insurers. Among the reasons for this, 'there are inadequate data to determine whether consistent global changes in climate variability or weather extremes have occurred,' although 'on regional scales there is clear evidence of changes in some extremes,' such as an increase in rainfall in the United States. As for the insurers' number one concern about risk in the future, 'knowledge is currently insufficient to say whether there will be any changes in the occurrence or geographic distribution of severe storms, e.g. tropical cyclones.'

But the insurers and bankers who convened to write the financial-services impacts chapters, led by Andrew Dlugolecki of General Accident, understood that the absence of certainty – and indeed 'knowledge' – is not the same as the absence of risk. 'Property insurance is vulnerable,' they concluded. 'Changes in climate variability and the risk of extreme events may be difficult to detect or predict, thus making it difficult for insurance companies to adjust premiums appropriately. If such difficulty leads to insolvency, companies may not be able to honour insurance contracts, which would economically weaken other sectors, such as banking.' Their conclusion was stark. 'There is a need for increased recognition by the financial sector that climate change is an issue which could affect its future at the national and international level. This could require institutional change.'

And so 1995 drew to a close. It was the hottest ever year in 136 years of records for land and at sea, fully 0.4°C above the 1961–1990 baseline. The hottest eight years on record had now all been in the last ten. Even in this ominous information there would be scope for carbon-club disinformation. The record of satellite-measured temperatures showed that 1995 was only the eighth warmest in 17 years of records. The carbon club would not be deterred by the fact that the record was so short, or for that matter that the technique averaged temperatures in the lowest six kilometres of the atmosphere, rather than sampling exclusively at the earth's surface.

The year 1995 also set new records for property-catastrophe losses. Total economic losses of $180 billion beat the previous record year, 1994, by a factor of three, and would have been a record even without the Kobe earthquake (a $100 billion loss). Insured losses of just under $16 billion were high, but lower than the record year of 1992's $28 billion and 1994's $17 billion. Munich Re registered 577 natural catastrophes worldwide. Among these, flooding in January and February in north-west Europe had seen 240,000 people evacuated in the Netherlands just 13 months after the Christmas floods of 1995. These floods involved $3.5 billion in economic losses and $900 million in insured losses. Swiss Re reported that if dykes on the Rhine and Meuse had been breached, the economic loss would have exceeded $50 billion.

The US National Oceanic and Atmospheric Administration reported 19 cyclones in the hurricane season, 11 of which had rated hurricane status, the second largest number in one season since 1886. Thankfully, these clocked up only $8 billion in economic losses, two-thirds of it insured. However, as Munich Re put it in their annual catastrophe report, this series of storms 'kept the population, the economy and the authorities on tenterhooks for almost five months'.

One of the main reasons for the large number of hurricanes, NOAA reported, was the cessation early in 1995 of the unusually long-lived El Niño. This raised the prospect that, if global warming was indeed responsible for intensifying the El Niño effect, as some climate scientists suspected, then perhaps the enhanced greenhouse effect would actually herald a *decrease* in Atlantic hurricanes.

Or then again, maybe not. As the Munich Re report observed, 'there are various arguments which imply that the climatically induced increase in ocean temperatures could in the long term lead to an increase in the intensity of tropical cyclones'. All this, the gloomy report concluded, would be part of a pattern if greenhouse-gas emissions were allowed to rise further and further.

JANUARY 1996, OXFORD

Greenpeace's senior management agreed to my part-time secondment to Oxford University during 1996, where I was to enjoy a one-year visiting fellowship at the Centre for Environmental Policy and Understanding. This institute was headed by Sir Crispin Tickell, environment advisor to UK Prime Minister John Major, and before him to Margaret Thatcher. Sir Crispin was widely credited with having been the main influence in converting Mrs Thatcher to the view that climate change was a major threat. A former senior British diplomat, he had written a highly prescient book on the climate issue over a decade before, at a time when few people – including scientists – understood the ultimate significance of an enhanced greenhouse effect.

In late January, Rolf Gerling – the man central to the success with the financial sector in Berlin – flew in his private aircraft to Oxford. I met him and Sir Crispin in the latter's panelled college office to hatch a strategy for building on what had happened in Berlin. Our idea was to bring together very senior people from banks and insurance companies with the chief executives of solar-energy companies. The plan was to try to catalyse a rolling process of marriage brokering between these two communities in the years ahead. A ground-breaking day introducing the two sectors to each other, we felt, might be the best way strategically to help get things moving. If we structured such a day well, and persuaded enough people to come, we ought to be able to make it clear that they had a common interest in abating climate change via clean energy. If so, we would have a chance of beginning to divert the river of carbon-directed capital into solar and other clean technologies. Our event was to be called, rather pretentiously, the Oxford Solar Investment Summit. We set the date for mid-December.

FEBRUARY 1996, NEWBURY, UK

The river of capital may still have been flowing unimpeded towards carbon, but change was in the air in a relevant area. In the UK, the Conservative government was locked into a massive roadbuilding programme. The Department of the Environment may have been producing worried noises about climate change, but the Department of Transport and the Treasury had other ideas, and more clout. Nobody could have foreseen the result. As the construction companies set to work, a remarkable new alliance emerged on the British political landscape during 1995 and early 1996. Middle England teamed up with the green and the disaffected and rose in vocal opposition to the new roads. TV news programmes showed tweed-jacketed ladies from the shires chained to bulldozers alongside young people with dreadlocks dressed in rags. Such counter-intuitive images had been appearing in British living rooms for months.

In mid-February, it became known that a huge demonstration would take place at Newbury, one of the most prominent sites of rural conflict between the roadbuilders and the protesters. Greenpeace, laudably taking a role away from the spotlight, organized buses from all over the country to ferry supporters down to the site. So it was that on a beautiful winter's day, I marched in a column of tens of thousands for 15 kilometres through the woodland and fields where the Newbury bypass was due to be built. I was surrounded by a cross-section of British society the like of which I had never expected to see unite on a single political issue. I saw a veteran soldier, a row of medals on his chest, explain to a TV camera why he had not fought for his country to see the countryside laid waste by greedy businesses and politicians. I saw middle-class families and their labradors walking and chatting with homeless travellers and their mongrels. I saw a dreadlocked tree-dweller, dangling from his simple hovel with high-tech mountaineering gear, shout to the thousands below 'your support means the world to us'.

And the British Government bent like a reed before the social fire they had kindled. Their U-turn on the roadbuilding programme was one of the biggest in the history of British Government policy. In Greenpeace, we could only stand back in awe, with a degree of envy.

What the oil and automobile companies were making of it, I could not imagine. The significance of these events, I knew, was not lost on them. Friends in the marketing consultancy world reported a heavy increase in requests for work.

It was in phenomena like this that the slow pace of change stayed just fast enough to keep hope alive within me.

MARCH 1996, NEW YORK

The third post-Berlin session of negotiations took place in Geneva in March, and finally it was a session at which governments achieved some progress. I missed it, being in New York to launch a book I had edited, and Gerling had published, based on the Berlin seminar. Germany proposed setting binding, across-the-board CO_2 reduction targets of 10 per cent from 1990 levels by 2005, and 15–20 per cent by 2010. This, the Germans professed, was the best target to start with: focusing on the main anthropogenic greenhouse gas, and the easiest when it came to assessing emissions. There were now three draft protocol proposals on the table: the long-standing AOSIS proposal of 20 per cent CO_2 cuts by 2005, a British proposal of 5–10 per cent by 2010, and the German proposal somewhere between the two.

The EU as a whole, while not yet having an agreed position on emissions in the short term – if that is what 2005–2010 could be called – made it clear, in partial support of Germany, that Europe supported action going beyond 'no regrets'. What this should mean, the EU suggested, was stabilization of atmospheric CO_2 concentrations at less than 550 parts per million – a doubling of atmospheric CO_2. This target should guide 'limitation and reduction efforts', which translated implicitly meant that the long-term goal would have to be deep cuts in emissions.

A scientific rationale for this 'no atmospheric doubling' target was crystallizing at the time. The Netherlands presented a new analysis of what constituted 'dangerous' climate change, and what it meant in terms of emissions to avert the danger. The maximum allowable threshold, the Dutch argued, should focus on 2°C as the ceiling for post-industrial rise in global average temperature. Beyond that, life would be simply too risky, the dangers of ruinous climatic disruption too great. Any emissions target in the short term must keep the world on track to remain below the 2°C ceiling. The Dutch called their new approach the 'safe-landing corridor' concept. The bottom line was that to keep within the safe-landing corridor, industrialized countries would have to achieve emissions cuts of 19–46 percent from 1990 levels by 2010.

The comprehensive Dutch study had a clear impact in Geneva. It and the first European target proposals were timely. The IPCC report was in the process of being digested by governments at this time, and at AGBM3 many of them conceded that impacts identified by the IPCC for doubling atmospheric CO_2 would be serious and even catastrophic. Yet, as IPCC Chairman Bert Bolin told the conference, on a business-as-usual emissions trajectory, CO_2 doubling would occur within 30 to 40 years.

And this was where the bad news began. It became clear from presentations made at AGBM3 that most industrialized countries would not succeed in their efforts to stabilize emissions at 1990 levels, as the Convention on Climate Change required them to. The convention's disappointed executive secretary, Michael Zammit-Cutajar, told a press conference that when it came to their obligations, most countries were heading in the wrong direction.

As far as Germany's ice-breaking proposal went, on the one hand some countries said they wanted all greenhouse gases included, but on the other, Australia led the naysayers with the counter-proposal that targets should be differential. Countries – such as theirs – with economies particularly dependent in some way on fossil fuels, they said, should be required to do less than those not so dependent.

Notwithstanding Australia's self-interest, attendees at AGBM3 reported that there was a new sense of urgency in the air. But it was clearly going to be a long and difficult haul to any common agreement on emissions reductions – much less something close to the 'safe-landing corridor' – by the time of the Third Conference of Parties late in 1997.

APRIL 1996, ZURICH

Rolf Gerling was not the only member of the billionaire businessman club trying to help make global warming a price-sensitive issue during 1996. Stephan Schmidheiny, Swiss industrialist and founder of the World Business Council for Sustainable Development, had compiled a comprehensive book about the river of capital flowing to unsustainable technologies. *Financing Change* was due to be unveiled in April at twin launch events in Zurich and Cologne. All concerned rather hoped that this book and the Gerling volume I edited would complement each other. They were each firsts-of-a-kind in terms of subject matter.

Financing Change was a group effort by the WBCSD in that it drew on the efforts of a working group on which many of the companies in that huge organization served. The WBCSD was an alliance of 120 companies – many of them among the world's largest – from 20 sectors of industry in no fewer than 35 countries. Prominent members included ABB, AT&T, Ciba-Geigy, Dow Chemical, DuPont, Hitachi, Mitsubishi, Monsanto, Sony and Volkswagen. The oil-company members included BP, Shell and Texaco. Financial-services companies included Gerling, NatWest, SBC and Yasuda Marine & Fire.

The politics of the organization, and the evolution of those politics, were a source of endless fascination to me. This interest dated back to the launch of the WBCSD's first book, at the Rio Earth Summit. I was in Rio as part of the Greenpeace delegation, which had targeted the WBCSD – or the BCSD as it then was – as something approaching the devil incarnate. This earlier book, written by Business Council figures, had been called *Changing Course*. It was, according to the Greenpeace analysis, largely corporate 'greenwash' – the environmentalists' term for the dressing up of bolt-on adjustments to the status quo as groundbreaking, planet-saving advances. The BCSD and Greenpeace locked horns in a public-relations battle, and I was in the heart of it.

I engaged in that struggle with reservations. I had met Stephan Schmidheiny in Japan during 1991. I talked to him at some length, and was convinced that he was sincere in his concern about the global environmental crisis and the need for industry to change. Indeed, in Japan he had made highly public statements on global warming which nobody in Greenpeace could have found fault with. I worked up a healthy respect for his aspiration to work pragmatically for change from within, and use his personal fortune in the service of that cause. There were, after all, few enough top industrialists prepared to concede that there was even an environmental problem, let alone attempt to do anything to help abate it.

But Schmidheiny's pronouncements on the need for meaningful action were not echoed, at that time, by many of the companies on the giant corporate alliance he had set up. The chapter on energy in *Changing Course* offered no concessions to the need for even freezing emissions of greenhouse gases. The lead author was Roger Rainbow of Shell, a man I had met and found to be utterly obstructive on the issue. I had the impression in 1992 that Schmidheiny had set something up which then projected a cutting-edge message which he cannot have been comfortable with on a personal level.

I was not short of contacts within the WBCSD who were willing to discuss the politics of the organization off the record. These included bankers, insurers and even oil-industry people. During the years that followed, the reports coming to me from such contacts showed that the member companies often fought bitterly within their various working groups, with opinion sometimes polarized. On no issue was this more so than climate change. There appeared to be at least two geopolitical tendencies. One was that the debate was most intense between financial institutions wanting change, and oil companies – led by Texaco – which resisted. Texaco's Clem Malin, my opponent from Berlin, featured prominently in the tales frustrated bankers and insurers told me about efforts to gain consensus within the organization. The other pattern, it seemed to me, was that the voices in favour of substantive change were gradually gaining the ascendancy.

When *Financing Change* appeared on my desk I could see at once that this latter impression was correct. The first chapter began with a quote from *Institutional Investor* magazine: 'Will financial markets soon be systematically rewarding environmentally successful companies while penalizing offenders? Some serious people think so.' The book then went on, chapter by chapter, to defend and develop that theme.

To my surprise, I was invited to speak at both launch events for the Schmidheiny book, as part of a panel of business leaders. The invitation came from Switzerland with no strings attached. I would be speaking for Greenpeace, and nobody asked to see the speech I would give ahead of the event. In Zurich, I would be sharing the platform with, among others, the CEO of the Swiss Bank Corporation. In Cologne, I would be there alongside the CEO of Gerling-Konzern, with Deutsche Bank represented at board level. Schmidheiny and his colleagues were placing considerable faith in the idea that things were changing so manifestly for the better in the WBCSD that even their old enemy Greenpeace would be able to bring itself to acknowledge the fact, and maybe even say something constructive.

I set about the task of testing that possibility. I drafted a speech which reflected my personal view, and sent it off to those who would have to rubber-stamp it. Most testing of all would be the ultimate guardian of corporate brand position within Greenpeace: the press desk at the German office. I would be speaking for the group in Cologne in front of several hundred business executives, with all the main German media outlets represented by economics correspondents.

To my relief, I got my sign-off with minimal editorial changes. The WBCSD was not the only evolving organization in town.

Had the Business Council gone deep green, or had Greenpeace gone soft on business, I asked in my speech. 'Neither, I think. The main reason we are here together today, I would suggest to you, is that where the global environmental crisis is concerned a new and intriguing whiff of hope has appeared in the air. Herr Schmidheiny and his colleagues call this a "sustainability reflex," and they believe that such a reflex is stirring with particular sense of purpose in the financial markets. In Greenpeace, we would agree with that. There is clear evidence that just at the time governments, for the most part, are running out of steam in their efforts to face up to the global environmental crisis, the financial institutions are beginning to show signs of serious concern.'

In his own speech, Schmidheiny said one thing in particular which stuck in my mind. The financial markets as they exist today, he argued, were not promoting sustainable development. In line with the book, he had recommendations for ways in which every part of the financial sector should change. The insurance industry, for example, had to think about investment. 'If insurers begin to invest only in enterprises they would be happy to insure, this could start a megatrend. That is my hope.'

His hope and mine alike. Later, after Kyoto, we would have the chance to work together on making that hope, and others like it, come alive.

MAY 1996, LONGBOAT KEY, FLORIDA

The conference suite in the luxury hotel was packed as Florida Insurance Commissioner Bill Nelson gave his speech. The setting was allegoric in the extreme. Outside the window an endless strip of sand stretched along the coast, a typical Florida beach along which I had run earlier. It separated the Gulf of Mexico from million-dollar residences by a matter of metres, for kilometre after kilometre. I now understood why, of the $2,000 billion of insured assets on US coasts, $1,000 billion were in Florida alone. And right by the hotel I had watched lorries delivering sand to the beach, where it was being ploughed into the eroding shoreface by bulldozers.

With this image in my mind, I listened to Nelson. The Commissioner had been an astronaut. He had been on a shuttle launched ten days before *Challenger* blew up. He described his ecstasy looking at earth's blue and white against the black void. He could define two phases of his life: before and after this experience. Similarly, he said, the people of Florida had two phases of life experience: before and after Hurricane Andrew. The insurance experience had been awful since Andrew, and the market had still not recovered. But people had to understand that, because of global warming, the worst crisis could yet lie ahead.

JUNE 1996, MILAN AND DENVER

With the Second Conference of Parties due for July in Geneva, and substantive discussion of emissions reductions on the agenda, it was vital for the carbon club to attempt to discredit the IPCC's Second Scientific Assessment. They began the process in April by publishing two reports. The first, from the George C. Marshall Institute, concluded that there was 'no evidence' that human activity caused global warming. The right-wing think-tank devoted only 22 pages to trying to prove this case, compared to the 572 of the IPCC's scientific assessment, considerations of size and authoritativeness rarely featured in media coverage of the war of claim and counter-claim. A second report featured the inevitable suggestion that the satellite record was more accurate than the ground-based temperature record. It was written by Patrick Michaels, one of the main dissenting scientists, and released not by the American Academy of Sciences but the Western Fuels Association, a prominent Global Climate Coalition member. Michaels was a flashy showman with whom I had debated for the first time at a climate-change conference in Washington the previous December. If not quite in the Richard Lindzen league, he was a man well able to convince those willing to be convinced, and I had no doubt that his report would cause the IPCC great problems in the USA.

The most serious attack did not come until June, when the IPCC's Second Scientific Assessment appeared in print. The GCC immediately released a nine-page analysis of the changes that Ben Santer had made in the chapter on detection. John Schlaes, GCC executive director, told the press that there had been 'substantial deletions and significant alterations' to the scientific material. He sent a letter of protest to Bert Bolin, IPCC Chairman, and copied it widely around the US Congress. 'These revisions raise very serious questions about whether the IPCC has compromised, or even lost, its scientific credibility. The changes are not just about grammar and punctuation. They go deeper and we want to know why they were made.'

Bert Bolin faxed a letter to the GCC, saying 'Your allegations are completely unfounded.' Ben Santer told the press that the allegations were 'dangerous and absurd'. The changes had been in response to peer review, to increase the scientific clarity. They were quite within the rules. Sir John Houghton, vice-chairman of the scientists' group, agreed. In a departure from his normal stand-back attitude to the politics, Sir John told *Nature* magazine that many of the revisions made in the IPCC-approved text had in fact been in response to political interference by the GCC, via proxy lobbying using Saudi Arabia and Kuwait.

But some of the mud dreamed up by the carbon club's lawyers stuck. *Nature* itself printed an editorial saying that 'the complaints are not entirely groundless'.

Santer, showing the first signs of a reaction that was to grow in the months ahead, said, 'I am really troubled by what is going on. This appears to be a skilful campaign to discredit the IPCC, me, and my reputation as a scientist.'

It was to become worse. On 12 June the *Wall Street Journal* published an article by Frederick Seitz, chairman of the George C. Marshall Institute. In one of the most sickening statements I have ever seen by a scientist, he called Santer's editorial changes a 'disturbing corruption of the peer-review process'.

The Global Climate Coalition egged the poisoned pudding by specifying that 'the changes quite clearly have the obvious political purpose of cleansing the underlying scientific report'.

'This is terrible what's going on, just terrible,' Santer told *Science* magazine.

You could smell the anguish behind the words of the carbon club's latest victim. When I had endured their unstoppable perversion of my 'worst-case' opinion survey, I at least knew that in the trenches of a war, dirty tricks happened. I had chosen a life in the trenches. But Ben Santer was an unassuming career scientist, and he hadn't.

In June, the World Renewable Energy Congress took place in Denver. One of the main themes, inevitably, was capital. The wind, biomass, hydropower, wave power, solar and other renewables industries sent delegates in their hundreds. I was invited to give the after-dinner speech at the congress banquet. I flew to Denver via Milan, where a European Union conference on renewable energy had just endorsed a statement exhorting governments to deliver a target of 15 per cent renewable energy by 2010. In Denver, I did my best to inspire the delegates with a vision of emerging opportunity. I was unashamedly partisan, interspersing the speech with tales of carbon-club perfidy, invoking the tormented spirit (or so I hypothesized) of John D. Rockefeller, and attempting to cast those who professed the inevitability of carbon's continuing hegemony as unimaginative and worse.

I flew on to California, then back to Michigan, for further work on the solar trail. I had plenty of time to think critically about my own rhetoric as I crossed America. Did I believe it 100 per cent? I knew I believed fully in the *potential* for a clean-energy explosion, but I didn't have such total confidence in the inevitability of it. I still couldn't envisage a *specific* set of circumstances capable of forcing the river of carbon-bound capital to break its banks. There was too much inertia and lack of imagination in the system, not to mention the malign intent of the vested interests.

But I schemed – and I suppose dreamed – on. I had resolved in the spring to leave Greenpeace after the Oxford Solar Investment Summit at the end of the year. From 1997 onwards, I wanted to fight to divert the river of capital from inside the solar market, pulling; not outside, pushing. I intended to cast fortune to the wind and try to set up a solar-energy company. I didn't yet quite know what the company would be or what it would do, but while I was in America I thought of a name for it. Whatever it ended up looking like, that company was going to hold true to a core value built around the greenhouse-abatement imperative, and its logical end result. Even the name should reflect it.

The Solar Century.

9

Cracks Appear in the Carbon Club

July 1996–April 1997

JULY 1996, GENEVA

As the Second Conference of Parties approached, Europe laid its stall out. On 25 June, a Council of EU environment ministers formalized firm upper limits for global average temperature rise and atmospheric CO_2 concentration, in line with the suggestions the EU delegation had made at AGBM3 in March. Their communiqué made historic reading. 'Given the serious risk . . . and particularly the high rate of change, the Council believes that global average temperatures should not exceed 2 degrees above pre-industrial level and that therefore concentration levels lower than 550 ppm CO_2 should guide global limitation and reduction efforts.'

Now, 550 ppm (parts per million) would mean a doubling of pre-industrial atmospheric CO_2 concentrations. The atmospheric concentration before fossil-fuel burning began was around 280 ppm. At the outset of the period chronicled in this book, it was 350 ppm. In July 1996, it was almost 360 ppm. If no efforts were made to cut emissions, the concentration would double within fifty years.

The ministers made it quite clear that they understood what keeping atmospheric CO_2 concentration below 550 ppm would mean: the reduction of global emissions to less than half the current level. Given this tough target, the ministers said, it would be essential to start the process of reductions at COP3, not merely to limit the growth of emissions.

Corporate America was also setting out its stall, but rather differently. On 8 July, the day on which a week-long pre-COP2 AGBM session began in Geneva, the White House received a 'Dear Mr

President' letter signed by 119 chief executives and chairmen, including those of Amoco, ARCO, Chevron, Chrysler, Exxon, Ford, General Motors, Mobil, Occidental, Shell, Texaco and a host of coal companies. The only interesting absentee was BP. The bottom line cannot have been encouraging to anyone on Clinton's staff thinking of the USA making a bold move in Geneva. 'The US should not agree to any of the three proposed protocols presently on the negotiating table. Your leadership on this issue is critical to assuring a continuing strong US economy.'

The next day, the insurance industry came to Geneva to offer a different view. The UNEP Insurance Initiative was almost as shrill as the Global Climate Coalition's standard fare. 'Human activity is already affecting climate on a global scale,' it read. There were 'serious implications for property insurers and reinsurers.' The advice to governments was stark. 'We insist that, in accordance with the precautionary principle, the negotiations . . . must achieve early, substantial reductions in greenhouse-gas emissions.'

This heady material was presented at a side meeting by a small group representing British, Swiss and Norwegian companies. But having presented the position statement and defended it in a short question-and-answer session, the insurers left. Dozens of the carbon club's foot soldiers would be there in the Palais des Nations spinning mischief for the full two weeks, of course. The insurance industry had parachuted in and disappeared, just as they had at the First Conference of Parties in Berlin the year before. This was not the way to influence the process, at least in a substantive way. In Berlin, a senior US government official had shaken his finger in my face and said of my insurance-industry friends, 'if these people aren't here full-time, they don't exist.' I had relayed that story countless times to insurers in the year that followed, but to no avail.

I arrived in Geneva after AGBM week, in time for the Conference of Parties proper, itself a week long, and to be attended by ministers. The speeches began on 17 July. A sizeable international press corps and hundreds of NGOs waited anxiously to hear whether the carbon club had shaken the faith of any governments in the IPCC Second Assessment. But as the day wore on it became clear that they hadn't.

The report was ringingly endorsed in almost every speech. Costa Rica, speaking on behalf of the full G77 group of developing countries and China, referred to the report as 'the most authoritative scientific assessment of climate change in existence'. Ireland, speaking for the EU, said the report left 'no room for doubt about the expected adverse effects of climate change'.

The British secretary of state for the environment, John Gummer, was in inspired form. He began his speech with a statement that ensured the attention of the listening hundreds. 'This very building ought to present us all with an awful warning. It is, after all, here that nations, full of good intentions, committed themselves to taking common action in the League of Nations – to prevent war. It was here, too, that we failed to take that common action; we failed to stand up to powerful interests; we failed to call the bluff of the purveyors of falsehood who put their selfish concerns before the interests of the world community. And today, we must not make those mistakes again. The credibility of the governments of the world is on the line.'

Gummer maintained this level of rhetorical edge throughout his short speech. He attacked Australia, now the clear front-runner in the foot-dragging camp. 'It's simply not good enough for major producers of fossil fuels, both oil and coal, to claim that their financial interests should stand in the way of progress in making significant reductions in greenhouse-gas emissions.' By the time he got to the end, most of the audience was rapt. 'The alarm bells ought to be ringing in every capital throughout the world,' Gummer concluded.

Everything turned on the attitude of the United States. If they shifted at the eleventh hour, just as they had in Berlin, then the second COP would end up with a product that left a chance of a protocol at the third COP, as the Berlin Mandate envisaged. We now knew that COP3 would be taking place in Japan, in the city of Kyoto. Delegates were already referring to the goal of the negotiations as 'the Kyoto Protocol'.

Undersecretary of State Tim Wirth began, as he had to, by addressing the carbon club's attack on the IPCC. 'The science calls us to take urgent action,' he said. This was clear, despite the work of the 'naysayers and special interests bent on belittling, attacking and obfuscating climate-change science. So let's take a false issue off the table:

there can be no question but that the findings meet the highest standards of scientific integrity.'

There was little room for doubt here. But what of the policy response? The Clinton Administration had been dragging their feet for months in the run-up to COP2. European frustrations, in particular, were running high. Indeed, there were many people in the Palais that day who expected that no new ground would be broken.

Tim Wirth duly announced that the USA was offering to set legally binding targets to reduce emissions of greenhouse gases beyond the year 2000. He did not specify what the American target would be.

There would be a real chance now of a Kyoto Protocol. The game was still on. The climate negotiations were still alive.

No sooner had Wirth finished than the press releases began to appear in piles around the Palais. The new US position, so the Global Climate Coalition said, could 'force Americans into second-class lifestyles'. John Schlaes announced that he was 'surprised and deeply concerned'. The environment groups, the UNEP Insurance Initiative, and the Business Council for a Sustainable Energy Future professed themselves happy to varying degrees.

The first signs of divisions now appeared in the GCC. One of the Edison Electric Institute's representatives was Charles Lindermann. He was there as a GCC member, and since the EEI was a creature of the electricity industry, he also represented dozens of US electric utilities. Lindermann refused to endorse the GCC's statement. He told the press that 'We sell electricity, we do not care where it comes from. The power lines do not know the difference if it comes from coal, a windmill or a solar cell. We are with the future, not the past. We know we cannot go back to the old days.'

It was a brave statement, and one which was guaranteed to get him into trouble back in Washington. But it gave hope to many people. The carbon club was showing its first sign of disunity, and after COP2 there would be more to come.

In the wake of the US shift in position, work on a draft ministerial declaration began in earnest. In a critical six-hour late-night session, China was reported to be playing a major role in hammering out a declaration based on the change of US position, along with the UK and Brazil.

The product circulated the following morning under tight wraps. It was rumoured to be easily good enough to give negotiators the necessary impetus for a successful run-in to Kyoto, and the end goal of a protocol with real targets and timetables.

Australia and Canada worked ardently during the morning to unravel the agreed text. For a long while now, a group of countries known as JUSCANZ had been coordinating their negotiating. Simply stated, these were the developed countries with reasons to want to slow down the European Union: Japan, the USA, Canada, Australia and New Zealand. Japan was wholly dependent on imported oil and coal, and already far less energy-intensive than other developed countries. The USA had its problematic heartland of coal- and oil-based opposition. Australia, with a government of Conservatives not in the John Gummer mould, was now brazenly defending its position as number one coal exporter. Canada seemed to have built its position around Albertan oil. New Zealand – a huge disappointment to environmentalists – seemed fixated on using its forests to avoid making CO_2 cuts, and was now the principal defender at the climate negotiations of the 'net approach' – allowing emissions credits for carbon soaked up by forests.

The change in the US position had cut the ground from beneath the JUSCANZ alliance, leaving Australia in particular dangerously exposed. To the extent that they could argue they were part of a blocking group, the Australian Government did have a chance of quelling domestic criticism. But isolated, they would be pilloried in their press, and risk turning the environment into an electorally harmful issue. Australian diplomats scurried around doing the dirty work of the Australian coal industry, showing little apparent shame that their predecessors had once been among the staunchest advocates of emissions reductions at the climate negotiations.

They worked to no avail. The two-page Geneva Declaration was presented to the plenary as a fait accompli. The Zimbabwean president of COP2 merely asked delegates to 'take note' of the document, a request that was greeted by loud, prolonged and doubtless relieved applause.

In their declaration, the ministers instructed diplomats 'to accelerate the negotiations on the text of a legally binding protocol or another legal instrument' by the time of COP3 in Kyoto. The desired product was set out in clear language: 'quantified legally binding objectives for emissions limitations and significant overall reductions within specified time frames'.

The key words here – going beyond the language of the Berlin Mandate – were 'legally binding' and 'significant'.

The only thing the Australians could do now was refuse to sign it, which they duly did, along with their neighbour New Zealand. They did so in the company only of Russia, and the predictable clutch of OPEC nations.

AUGUST 1996, HARTFORD, CONNECTICUT

The Clinton Administration came under immediate domestic fire after the breakthrough in Geneva. The presidential election campaign was now in full swing, and the Republican platform echoed the GCC hobby horses exactly: 'Republicans deplore the arbitrary and premature abandonment of the previous policy of voluntary reductions of greenhouse-gas emissions. We further deplore ceding US sovereignty on environmental issues to international bureaucrats and our foreign economic competitors.'

On 26 August, writing in the *National Underwriter*, Frank Nutter renewed his brave effort to persuade the US insurance industry, and no doubt the many Republicans at the top of it, that measures to reduce emissions would in fact not be misdirected. 'Global warming is not just hot air,' he wrote. 'The paradigm has changed. Will we only do the right things after we have exhausted all other possibilities? Or will we take the steps that are in our ultimate self-interest?'

Nutter's was still a lone voice. His industry was proving bewilderingly difficult to persuade. And yet elsewhere in the USA during 1996, there were signs that people and organizations were awakening to the greenhouse threat in a way they hadn't in 1995. Perhaps most notably, January's *Newsweek* ran a fascinating front cover. At the time, devastating blizzards had been bringing much of the east coast to a skidding halt. The cover showed a deserted and ice-blasted New York street. 'THE HOT ZONE' announced the headline. 'Blizzards, floods and dead butterflies: blame it all on global warming.' Here was understanding at last in a popular American news outlet: we were talking, after all, about *climate change* in response to global overheating. An editor had seen beyond the 'warming' tag. *Newsweek* was not alone, however. Reporting of the issue improved noticeably in 1996.

Three days after Frank Nutter's article appeared, I visited one of America's biggest insurance companies, in Hartford, Connecticut. As I walked into the office of my host, he was looking at a computer screen on his desk. On it was a satellite image of the north and central Atlantic. We stared together at three huge swirls of white marching equidistantly in a straight line from the Azores to the Florida Coast.

The first two of these swirls were named Edouarde and Fran. Each of them had the potential for ruin, from Dade County to Cape Cod. Not to mention great companies like the one I was visiting.

Edouarde spent much of a weekend seemingly aimed at Long Island. I had to be in Washington the next week. I fled Hurricane Edouarde, taking an early train to New York as the storm roared down on New England. Fortunately, the storm took a few turns to miss first Connecticut, then Massachusetts, then all but a corner of Maine.

That left Hurricane Fran. She hit the beach at Cape Fear in the Carolinas on 5 September while I was in Washington, the first major storm of the US season to do so. Thankfully, Fran was only Category 3 at the time, but she caused damage to thousands of homes in the Carolinas, Virginia, Maryland, Pennsylvania and Ohio. The US Property Claims Services Division made an initial estimate of $1.6 billion in insured losses, making Fran the tenth biggest weather-related loss ever, and bringing the number of all-time billion-dollar weather cats to 20. Fifteen of these had now been since 1987.

OCTOBER 1996, LONDON

In October, BP quietly let it be known that it was quitting the Global Climate Coalition. Telephone lines were buzzing with the news days before the story appeared in the *Financial Times*. Managing Director Rodney Chase explained that the British oil giant wanted more leeway to tackle the problem of climate change, and felt that 'precautionary action' was now appropriate.

At a conference in London, a senior BP official – not my debating partner, Klaus Kohlhaus – confided to me that the reasons went further: 'embarrassment,' he said, at the extremes to which the GCC had sunk in trying to discredit IPCC science over recent months. It made sense. Environmentalists at COP2 had noted GCC lobbyists enthusiastically applauding a speech by Saudi Arabia in which the OPEC country attacked the IPCC and the Geneva Declaration text. They noted at the time that Klaus Kohlhaus had looked less than impressed.

BP were not alone in losing patience with the GCC. The Arizona Public Service became the first utility to quit. Their chief executive told the press that 'global climate change is a serious problem and we need to take steps to deal with it'. This was a defection just as encouraging as BP's. A major study by the Natural Resources Defense Council had a few weeks earlier ranked all US electric utilities according to their exposure to potential liabilities from measures to cut greenhouse-gas emissions. The NRDC based its analysis on reported emissions of carbon dioxide per dollar of electricity revenue in 1995. The total exposure of electric generation to carbon dioxide emissions limits or taxes could exceed $60 billion annually, according to one calculation. 'The lower-risk utilities can and should take far more commercial advantage of their relatively low emissions,' said NRDC's Ralph Cavanagh.

As Mark Mansley had pointed out in the Delphi Report two years earlier – and ahead of its time – the investment implications of overdependence on coal were now becoming real. Real-life giant utilities were now conceding the case. Of 119 US utilities covered in the NRDC study, those with the lowest financial risk from anticipated controls on greenhouse-gas emissions were Bonneville Power Administration and Pacific Gas & Electric. NRDC was joined in the release of the report by BPA and PG&E executives. 'We are gratified but not surprised to find ourselves at the top of this list,' BPA administrator Randy Hardy said. 'In our power marketing, we already have begun to emphasize our unique lack of exposure to the risks associated with fossil-fueled generation.'

The churches did their bit to turn the investment screws, meanwhile. The Interfaith Center for Corporate Responsibility asked shareholders in companies who were members of the Global Climate Coalition to withdraw, arguing that 'in no case should (a member) company allow its name to be associated with any additional GCC activities that distort the facts'.

NOVEMBER 1996, TOKYO

Much buoyed by these emerging cracks in the carbon club's machine, I flew to Tokyo with Bill Hare for a week in the service of our colleague Yasuko Matsumoto. The formidable Yasuko had decreed that a vital component of success in Kyoto would be convincing Japanese industry that a protocol with emissions reductions was a good idea. At best, she argued, we had to play our part in persuading them not to oppose such a deal, even if they couldn't bring themselves actively to support it.

Yasuko had arranged a briefing by Bill and me to which over a hundred Japanese companies, including many majors in the financial-services, energy and construction sectors, had been invited. To everyone's amazement, almost all of them accepted, many sending extremely senior people. Never in the short history of Japanese environmentalism had such importance been attached to the perspective of an environmental group.

Before the day of that performance, Yasuko took Bill and me to see the environment minister, Sukio Iwatare. He told us that he was cautiously optimistic that Japanese industry could be persuaded to support emissions limitations during the run-up to the Kyoto Climate Summit. His agency was working hard on that, and he hoped we could play our part as well. There were hopeful signs already. The Chairman of Keidanren – the Japan Federation of Economic Organizations – had sent a request to a hundred business associations in the Federation to submit environmental action plans by mid-November, including measures to combat global warming.

On the day of the seminar I found myself nervous in a way I hadn't been for years. Yasuko and her colleagues had taken all the viewgraphs Bill and I used, and translated them into Japanese. They had organized for simultaneous translation to be fed into radio earpieces for delegates. The whole event was professional in the extreme, and Bill and I now had to repay the utter faith that had been shown in us. His job was to cover the science and policy, mine the business implications.

Afterwards, my Greenpeace Japan colleagues watched anxiously as journalists interviewed departing delegates. Bill and I in turn nervously watched our colleagues, waiting for feedback.

To our relief and joy, it was almost uniformly good. One industrialist had even said he wished the Ministry of International Trade and Industry would give briefings as comprehensive and useful.

A happy international party repaired that night for sushi and karaoke.

On 5 November, Bill Clinton was re-elected President of the United States. Barely hours after Clinton's acceptance speech at Little Rock, Tim Wirth was in Canada promising that the environment would be a top priority for the new administration, alongside health and education. Among the environmental priorities, Wirth said, global warming was key.

For the Canadian Government, none of this was particularly welcome. At the same meeting, Environment Minister Sergio Marchi did not seek to hide Canada's poor record to date. Like all industrialized countries, Canada pledged to freeze greenhouse-gas emissions by the year 2000 when it signed the Convention on Climate Change. 'It would be nice if I could tell you Canada is succeeding,' Marchi observed. 'The truth is, Canada is not doing as well as it should. Period. No excuses.'

Canada of course was a mainstay of the JUSCANZ alliance at the climate negotiations. So too was Australia. The change of heart in

Washington had left both these countries exposed, and on 22 November, in one of his first second-term speeches, Bill Clinton delivered a coded message to Australia. Standing in a park in Queensland, the president said: 'If present trends continue, there is a real risk that sometime in the next century, parts of this very park could disappear, submerged by a rising ocean. That is why today, from this remarkable place, I call upon the community of nations to agree to legally binding commitments to fight climate change. We must stand together against the threat of global warming. A greenhouse may be a good place to raise plants; it is no place to nurture our children. And we can avoid dangerous global warming if we begin today and if we begin together.'

DECEMBER 1996, OXFORD

As Christmas approached, the day arrived for the event I had been planning all year. The sign-up process had benefited from a snowball effect. The location had no doubt been a draw card, and the University's Environmental Change Unit had helped us to organize the event. Once it became known that chief executives and chairmen of insurance companies were intending to come, it became progressively easier to persuade senior people from other insurance companies and banks to participate. With this dynamic in play, the solar companies had queued up to send delegates.

On the day of the Oxford Solar Investment Summit, 80 people from more than 50 companies, institutions and organizations gathered in the ballroom of Oxford's Randolph Hotel. Most of the people under the chandeliers were chairmen, chief executives or executive directors. Roughly equal numbers came from financial institutions, solar companies and potential or actual solar consumer entities: the three corners of what I called 'the triangle of common interest'. The rationale for the gathering had been made clear to the participating companies from the outset, and in accepting invitations they had tacitly agreed to it: a collective desire to see if there were ways to accelerate the commercialization of solar photovoltaic energy, through profitable investments and consumer alliance, as a key strategic route to reducing the risks of climate change.

My core aim was to generate a sense of collective excitement about the potential for solar energy. The key to being able to achieve that would be the ability of the solar-energy industry and its advocates on the day to make a convincing case that theirs was a business poised to explode into huge new markets. To set us on that course, I elected to ask three of the biggest names in the solar PV business to kick the day off with an opening blast of a few minutes each.

Bob Kelly, chief executive of Amoco-Enron Solar Power Development Corporation, began. He gave a synopsis of the upbeat analysis I had seen him give several times before. It was the best possible start: a senior executive from America's biggest gas company talking the same language as the most enthusiastic environmentalist. Next came Wayne Gould, manager of distributive generation for Southern California Edison. This was a gamble. I had never heard Gould speak, I only knew of him by reputation. He came from one of the biggest electric utilities in the world – one that had something of a record on renewable energy, but was still very much fuelled by all that was old and traditional. Gould, however, was a revelation, as my advisors had told me he would be. He was born and bred a utility man, he told the gathering. His grandfather had been a utility boss in the days of coal, and his father a utility boss in the days of nuclear. The gathering might think he was now here as a utility man to tell them about how the future was in gas. He wasn't. He was here to tell them it lay in solar energy.

PV has made the grid system obsolete, Gould announced. People may think that sounded like a counter-intuitive opinion for a utility man to hold. But just as the dish had made the cable obsolete, so would PV make wires obsolete. Solar PV would inevitably gain primacy over other forms of energy: it was economically viable in many markets today, and would become economic in all soon.

The next presenter, Joachim Benemann, president of Pilkington Solar International, built on this perfectly. He described how PV technology was practical in northern latitudes of developed countries, particularly for integrated systems where PV material can be used to replace conventional building materials such as roof shingles, stone facings, glass curtain-wall areas and even new tiles. The paradox, he explained, is that whereas people do not concentrate on the cost of a normal building façade, once PV is proposed, the first question is usually about cost. Yet the costs are not much higher than for steel, glass and other materials in common use. Benemann showed a series of beautiful slides of projects his company was involved in, mostly in Germany.

This opening session had gone as well as I could have hoped. In the first coffee break, comments to me by insurers and bankers confirmed it.

'I had no idea it is so close to happening,' one investment manager said to me.

The 80 delegates spent most of the rest of the day in working groups, pondering how best to break solar PV out of its price trap, and into markets capable of attracting institutional investors in growing numbers. At the end of the afternoon, the groups reported back to a plenary session. I had asked investment people in banking and insurance to be the rapporteurs of the groups, and to be ruthlessly realistic in reporting, so that we had a reality check on how far we were from being able to help the solar market to take off. With the solar industry's advocates reporting, we would not be guaranteed that.

I listened to them report, as I know many people did that afternoon, with a rich mix of excitement and hope.

Sir Crispin Tickell, who together with Rolf Gerling had navigated the difficult task of chairing the two plenary discussions, now had the job of drafting a conference declaration based on the reports. Among the points embraced by the consensus were the following: 'The enhanced greenhouse effect presents us with a major global problem. Yet technologies exist that can mitigate that threat. One of them, solar photovoltaic (PV), is of particular interest because of its appeal to vast markets into which it can expand. It represents a major opportunity to reduce our current dependence on fossil fuels in the near to medium term, thereby adding to security of supply. It also represents the energy-generation system most friendly to the environment with a limitless future.'

The declaration went on to talk about the price trap. 'A major challenge is to bring PV to high-volume production, at which economies of scale would produce lower competitive prices. At a time when electricity consumers will have a wider choice of supply, markets seem poised for rapid growth with investment to match.'

Sir Crispin read the declaration to delegates at a cocktail party in the panelled dining rooms of Brasenose College, where the summit dinner was to be served under festive decorations and the unsmiling portraits of a legion of old college masters. Prolonged applause when he finished showed how well he had captured the mood of the day. As we ate dinner, I looked up at all the old worthies in the flickering candlelight. If I squinted, ignoring the sharp suits of the financial- and energy-industry people, the scene would have been little changed from any over the centuries. Yet the world had changed dramatically, and changed and changed again, during the tenure of the old masters hanging up there on the oak panels.

The Oxford Solar Investment Summit had shown, I hoped, that the world could change dramatically once more. True, there was a certain tone of 'come back to us when the solar revolution has started' from financiers – rather than 'how can we help you kick it off?' But then there was a major paradox inherent in multibillion-dollar financial giants looking for good investments in an embryonic industry in which companies measured their turnovers in mere millions.

There had to be a way. If only we could figure out a means to make the mix of solar power and capital achieve a critical mass.

As of 1 January 1997, in a new organization supported by Rolf Gerling and others in the financial-industry vanguard, born that December day in Oxford, my mission would be to find a way to help that happen.

10

A Crime Against Humanity

May–November 1997

MARCH 1997, WASHINGTON, DC

The European Union had for a long time now been the most progressive force at the climate negotiations, outside the Alliance of Small Island States. If there were to be any prospect of a protocol with cuts in Kyoto, the EU would have to make the first move. They did so on 3 March in Bonn, at the sixth AGBM session, while I was in Washington talking solar energy. Environment ministers from the 15 nations agreed that day in Brussels to a Union-wide position. They called for developed countries to sign up to a flat-rate reduction of 15 per cent in emissions of carbon dioxide, methane and nitrous oxide by 2010. This target should be the first step towards limiting the increase in global average temperature above pre-industrial levels to 2°C.

AOSIS and environmental NGOs had been vehemently advocating a 2005 goal as essential in order to ensure early action to prevent a dangerously high rate of climate change. This argument had now hit the rocks. The Dutch, who then held the EU presidency, had supported a 2005 target; the UK and France had opposed it. The 2010 target was a compromise.

None the less, the USA, Japan and Australia opposed the EU's suggestion immediately. The EU's 15 per cent figure was a composite, Union-wide sum. Within the EU big cuts would be made by some members, such as Germany and Denmark, who would both take on 25 per cent; meanwhile, actual increases would be permitted by others, such as Sweden, Spain, Ireland, Greece and Portugal. In Portugal's case, the increase would be as much as 40 per cent. This idea was to become known as the 'EU bubble' proposal.

The JUSCANZ countries were openly critical. Such a differentiation of targets within the EU was not acceptable, they said, when the EU opposed differentiated targets among nations outside the Union. Australia and Japan called the EU stance hypocritical.

Other countries and environment groups criticized the EU for having only three gases in its bubble. The long-lived gases – HFCs, PFCs (perfluorocarbons) and SF_6 (sulphur hexafluoride) – should also be included. They were currently being produced only in small quantities, but they held the potential for strong emissions growth in the future, unless regulated.

Further divisions emerged on the subject of targets for developing countries. The USA wanted the developing countries to agree to targets of their own. The EU disagreed. Developed countries should lead the way, they said, and indeed had already agreed to do so in the Berlin Mandate itself.

The talks adjourned in disarray. Many countries had now proposed limits for emissions reductions, and even draft protocol language. But, as one observer put it, this meeting simply catalogued them.

The Global Climate Coalition issued a statement at the end of the negotiations saying that the lack of marked progress in narrowing differences was not surprising, but neither was it necessarily bad. They voiced doubt that the EU would be able to meet its emissions target of 15 per cent below 1990 levels. 'That's a negotiating target, not a commitment,' said Constance Holmes, chair of the GCC's operating committee. 'There is no analysis provided to demonstrate that it is attainable.'

John Schlaes mixed an edge of triumphalism with his usual grotesque hyperbole about the costs and effects of action. 'The bullet train to Kyoto has slowed to a crawl, and for good reason,' said the GCC's executive director. 'Maybe we'll be spared a rush to a judgement that could be inordinately costly to everyone and produces no environmental benefit.'

As though this portrayal was not depressing enough, I had talked to an informer inside the GCC while I was in Washington, someone who was uncomfortable with the growing dishonesty of his umbrella group. John Schlaes was about to be fired, he told me. For an instant,

I had been elated. But then I heard the reason. The GCC's paymasters did not consider him sufficiently hard-line.

The picture of economic pain as a result of emissions reductions that the GCC had been embroidering was beginning to look threadbare. In mid-February, a remarkable document added itself to the long list of declarations on the consequences of global warming. More than two thousand economists, running the gamut from conservative to liberal, published a consensus statement professing that the USA could reduce its emissions without damaging its economy. The statement called for new measures such as carbon taxes and trading of marketable emissions permits between countries. It asserted that such measures could be enacted without harming American living standards, declaring that well-designed policies relying on market mechanisms may in fact improve US productivity in the longer run.

As for the economic costs of inaction, the reverse was true. In late March, Munich Re published its annual analysis of the global catastrophe damage bill. The record showed that nature had been kind to the global insurance industry in 1996. The industry suffered $9 billion in insured losses in 1996, well below the figures for 1995 and 1994. As Munich Re put it wryly, we had finally had a 'normal year'. But, warned the giant reinsurer, this did not mean there was less cause for concern. The general trend towards ever-increasing numbers of catastrophes was continuing. The threat to cities was particularly severe, even potentially apocalyptic. Reading the report, I came across a sentence which I knew must have been inserted only after some editorial consternation. 'According to current estimates,' the report concluded, 'the possible extent of losses caused by extreme natural catastrophes in one of the world's major metropolises or industrial centres would be so great as to cause the collapse of entire countries' economic systems and could even bring about the collapse of the world's financial markets.'

I had by this time read and heard references to the potential bankruptcy of the global insurance industry on many occasions, but this was the first time I had seen a reference to ripple effects from an insurance collapse generating a meltdown in the capital markets.

Yet, although it clearly understood how high the stakes were, bewilderingly, the world's biggest reinsurer had still not joined the UNEP Insurance Initiative to lobby for progress at the climate negotiations.

APRIL 1997, CORAL GABLES, FLORIDA

One American reinsurer, Employers Re, had by now joined the UNEP initiative. In late April the new member staged a conference on climate change at which every speaker known to be concerned about global warming was matched with a sceptic or a carbon-club luminary. The reason for this, according to well-placed sources, was that a problem had arisen with Employers Re's shareholders. The company controlling the reinsurer, GE Capital, was also a well-known bankroller of steam turbines of all descriptions. It seemed that GE Capital had leant on the organizers to make sure that the alternative view of climate change was 'fairly' represented.

Among the speakers at the conference in Florida was a face new to me. Ross Gelbspan, a Pulitzer Prize winning journalist who had worked for the *Boston Globe* and the *Washington Post*, had been researching the carbon club's dissidents for more than three years. He had written a book on them, and it was soon to be published. I knew at once that I had found a valuable ally. Gelbspan's research had brought to light some fascinating material on the relationships between the carbon club and their scientific hit men. All the main dissidents had testified under oath at a Minnesota hearing in 1995 about the effects of burning coal. In the course of this, they had each admitted to taking consultancies with oil and coal interests amounting to hundreds of thousands of dollars. Patrick Michaels' paymasters included Cyprus Minerals, the largest single funder of the rabidly anti-environmentalist Wise Use movement. Robert Balling's included Kuwaiti government sources.

In the audience all day in Florida was an executive from GE Capital. I talked to him over coffee. He was a tight-lipped individual with sweat on his brow, and it was clear he knew little or nothing about global climate change. He was there, I surmised, to watch Employers Re's performance and report back to the shareholders.

A few months later, Employers Re withdrew from the UNEP Insurance Initiative. The admirable Frank Nutter would still be a lone voice in his industry, four years after he had first spoken out.

JUNE 1997, DENVER / NEW YORK

Elsewhere in the business community, others were less slow to embrace change. On 19 May came a remarkable development. Speaking to an audience at Stanford University, BP chief executive John Browne announced that BP had decided that global warming was now a real threat, requiring a meaningful response. From now on, his company would as a consequence be taking solar power very seriously. BP planned to increase its sales of solar PV from $100 million a year today to $1 billion a year over the next ten years.

I was elated. I knew at once that this would send a signal into the markets that would make my work of trying to persuade investors to back solar much less difficult.

Greenpeace, however, attacked BP's change of heart for not going far enough. The company's aggressive development of new oilfields in the Arctic and the Atlantic had not been in any way affected. The climate crisis was too severe for any new oilfield development. They were right, of course: BP did want to have its cake and eat it. Browne had made it clear that solar energy could only ever be additional to oil and gas, never an alternative to it. By having the goal of increasing its sales tenfold in ten years, BP Solar was buying into only a slight acceleration of business-as-usual growth for the industry.

None the less, this speech was a major landmark. The oil industry was now comprehensively divided, and this change of heart by such a large industry player would clearly stand a chance of boosting the will of governments to act at Kyoto.

America's corporations seemed oblivious to this change of course by BP. On 10 June, 130 chief executives took out an advert in the

Washington Post spanning three full pages. The first page asked, in huge letters, 'HOW DO YOU PROTECT AN EARTH IN THE BALANCE?' Spread across the next two pages was an answer of sorts: 'WITH A BALANCED APPROACH'. Above their signatures, the business chiefs explained that the issue of whether or not the Clinton/Gore Administration signed the Kyoto treaty 'may be the most important economic decision of this century and the next as well'. The industry leaders included all the usual oil-and-coal suspects, but many more besides. 'A balanced approach,' they said, 'is only possible with careful study, input from a wide variety of sources, and extensive public debate. We strongly urge the Clinton/Gore administration not to rush to policy commitments until the environmental benefits and economic consequences of the treaty proposals have been thoroughly analyzed.'

This gambit, thick with the disingenuous codewords the carbon club had used throughout the climate negotiations, set new standards in overt corporate pressure on government. Every country had its companies lost in scepticism about climate change. But in the USA the scale of the collective denial was unique. There was something primitive, even frightening about it.

The play on his book by such a huge cross-section of corporate America must have hurt Al Gore. As bad as this was for the White House, the corporate laggards had plentiful supporters in Congress. Shortly before the Denver summit, Senator Robert Byrd, a West Virginia Democrat, and Chuck Hagel, a Nebraska Republican, co-sponsored a Senate resolution that would rule out mandatory emissions limits if developing countries refused to adopt 'meaningful' targets themselves. This moved the most basic of goalposts in the convention: that because the developed countries had caused most of the greenhouse emissions in the past, they should act first to reduce them.

The key argument offered by Byrd and his supporters was that if America took on emissions commitments while the likes of China and India didn't, industries and therefore jobs would be sucked out of the USA and into the developing countries. It was an image designed to inflame sensibilities, but was based on a near nonsense. Two-thirds of US carbon dioxide emissions were in the building – domestic and commercial – and transportation sectors. These were not in any way amenable to export. The remaining third came from industry, where the competitiveness issue was limited to a very few energy-intensive activities, representing perhaps just a few per cent of manufacturing shipments and jobs, and even these could be offered a measure of protection with efficiency improvements.

US environmental groups watched in horror as the senators, none the less, lined up to sign Byrd's resolution. By the time of the 1997 G7 summit in Denver, 61 out of 100 had done so.

It seemed clear to me that the White House had to take a lot of the blame. If they had come out fighting long ago, if not at the start, they could have made the arguments count. The history of the global warming issue tended to suggest very strongly that if people were exposed to the arguments for long enough, they tended to be convinced. I was not alone in taking this view. So distressed were US environmental groups with the administration's lack of fight, that some of them were now warning that green voters might abandon Al Gore in the Democratic primaries of the year 2000 in favour of Richard Gephardt. The Sierra Club, a well-known US environmental group, had begun taking out TV adverts in selected primary states exhorting the administration to 'stand up to the special interests'.

But at the Denver summit, as I watched with a large contingent of aghast representatives from environmental groups and clean-energy businesses, all the standing-up the administration did was to the other G7 countries. President Clinton refused to make a specific commitment on US targets for greenhouse-gas emissions, and his officials insisted that urgent language on the issue be excised from the summit communiqué.

Immediately after Denver came another summit: a special session of the UN in New York to review progress five years on from the Rio Earth Summit. This was a genuine environmental summit, with not an economics correspondent in sight. Sixty world leaders would be in town, with diplomats from 200 countries, and thousands of NGOs. Whatever happened here would bear significantly on Kyoto's prospects.

Tony Blair, the new UK prime minister, was one of the first world leaders to speak at the Rio-Plus-Five summit. He set exactly the right precedent. In one of the strongest speeches I have ever heard a leader make on the environment, he made it transparently clear that the collective record to date had been unacceptable. He spoke not just as a prime minister, he said at the outset, but as a father. The EU had proposed a new and challenging target for greenhouse-gas reductions, he continued. 'We in Europe have put our cards on the table. It is time for the special pleading to stop and for others to follow suit. If we fail at Kyoto, we fail our children, because the consequences will be felt in their lifetime. And we must all deliver on the commitments we make. Setting new targets means little if old ones are ignored.'

After Denver, nobody expected President Clinton to announce specific commitments in New York, notwithstanding the attention of the world media on the event. He duly didn't. But the president did announce two new initiatives designed to head off the welter of criticism from the European leaders. The first was a billion-dollar package of aid for developing countries to help them reduce greenhouse-gas emissions. The second was a programme to install solar panels on a million roofs in the USA by the year 2010.

For many, hope was running out fast as a result of the five years of failure since Rio. The island states came top of this list. In New York, the AOSIS heads of state voiced exactly the same fears and frustrations that they had in Rio. The difference now was that the islands were experiencing sickening foretastes of their future. On Kiribati, for example, two tidal surges had rolled across Tarawa Atoll in January and February. I had read the accounts of these with horror. They came without rain or storm: unrelenting rising tides lapping ever higher, all the more sinister for the stillness of the day as they swallowed up homes. Nobody on the atoll had seen anything like it before.

JULY–OCTOBER 1997, OXFORD

On 24 July the Clinton Administration made a belated effort to win some public support for greenhouse-gas cuts. The president lined up some of the USA's foremost scientists at the White House to explain the climate change problem to a bank of TV cameras and the nation's press. 'There is ample evidence that human action is already disrupting the climate,' he said. 'We can see the train coming but most ordinary Americans, in their day-to-day lives, can't hear the whistle blowing.' The president promised that the occasion would be the beginning of a sustained effort to win US public support for action in the run-up to Kyoto. 'I believe it will give us the support we need to take the action we are going to take,' he said.

Rumours in Washington had it that Vice-President Gore, stung by recent events, had finally persuaded the president to give the issue high priority. Bill Clinton, it seemed, had listened.

This ought to have been encouraging, but on the very next day the Senate delivered a hammer blow. Support for the Byrd Resolution had proved to be not just bipartisan, but unanimous. The final count was an incredible 95–0. Every US Senator now supported the call for the administration to avoid signing any Kyoto agreement that did not entail emissions commitments for developing countries.

Said Republican co-sponsor Chuck Hagel, 'The Byrd–Hagel resolution is a complete rejection of the Berlin Mandate.'

The administration had put in some work trying to amend the resolution while it was still under discussion in the Senate, but had failed. Now the White House did what it could to minimize the damage. The chosen course was to pretend that nothing much had changed. 'The notion that developing countries need to be part of the solution is something we agree with,' said Katie McGinty, head of the president's environment office. 'This strengthens our hand,' said Tim Wirth, leader of the US delegation which had negotiated the Berlin Mandate.

This was transparent. Wirth's comment sat beside Hagel's in complete discordance. The carbon club's victory could not be hidden.

On the eve of the seventh session of the post-Berlin phase of climate negotiations, in late July, the river Oder burst its banks causing a 'once in a thousand years' flood in Central Europe. Much of Poland, and large areas of the Czech Republic and eastern Germany, were under water. Munich Re conservatively estimated at least $500 million in damage. Once again, Europe watched troops deployed to try to prevent disaster from turning into catastrophe. Thousands fled the region of Germany known as Berlin's back garden, hoping that protective dykes would stop their homes and farms from being destroyed. The largest military deployment for thirty years toiled to keep the dykes from breaking.

Fortunately, they held. Insured losses were light, flood exclusions being widespread. But economic losses were bad enough to threaten the process of economic recovery in Poland and the former East Germany. Another foretaste of the future had featured for a few days in the news columns, then to disappear.

I could not go to the seventh session of talks. My efforts to find investments for the solar industry were taking up all my time. Early on in the session, so the NGO newspaper *Eco* reported, Don Pearlman was overheard talking to a Nigerian delegate. Gleefully he chuckled that if developing countries were to dig in and react to the Senate resolution, insisting on no new commitments, then the Kyoto protocol would be dead in the water. 'We can kill this thing,' Pearlman told the OPEC man.

The carbon club had several dozen people in Bonn to help Pearlman hit that target, 25 on the Global Climate Coalition's delegation alone. They were armed with copies of the Byrd–Hagel resolution translated into six languages. Their strategy was simple: to inflame sensibilities so that the negotiations would collapse from within. Congressional staffers were in Bonn for this session as part of the official US delegation. They too were overheard by NGOs openly discussing their wish to kill the whole process.

It now looked as though they had every chance. The USA still refused to announce a commitment target, other than to say that the AOSIS target year of 2005 was out of the question. Japan had no target proposals to table either. Worse, both the USA and Japan attacked the EU bubble, fostering transatlantic acrimony to add to the North–South dimension. Negotiators banned NGOs – industry and environmental lobbyists alike – from their sessions on reduction targets, ostensibly to speed progress. But it was impossible to see how they could make progress, or even negotiate at all, when two of the key players had not even tabled proposals.

Meanwhile, in other contact sessions the USA pushed hard on the advantages of including emissions trading and joint implementation in a protocol, this time supported by Japan. Trading emissions quotas meant that those who cut more than their target could sell quotas to those struggling to meet targets. Joint implementation meant that countries which invested in emissions reductions overseas could claim the reductions for credit at home. On these issues the EU had long been reluctant, but now professed itself willing to compromise, provided the emissions reduction target in the protocol was adequate. Other signs of compromise were in the air. More and more countries now favoured the 'comprehensive' approach – targets for a number of greenhouse gases, not just carbon dioxide, or gas-by-gas targets. Similarly, net accounting seemed to be enjoying a growing tacit acceptance.

The key issue remained the targets. Even if the White House had intended to table targets here, the carbon club's success in the Senate made it very difficult for them to do so. Meanwhile, they had to be seen to be taking steps to include the developing countries in the target-setting. At the very least they had to try to persuade the key countries – China, India and a few others – that some commitments that could be seen as 'meaningful' might be a good idea. 'Meaningful' – the key word in the Senate resolution – seemed to be something the Clinton team would try to define later, in their ratification battle with Congress, provided some new concession could be made on targets now, and packaged as a developing-country target. They developed a codeword for the task: 'evolution' of the Berlin Mandate.

China, India and the rest of the G77 were predictably in no mood to see a key principle they had negotiated in both 1992 and 1995 'evolved' in any way at all. The only developing country showing any patience with the Clinton Administration's dilemma seemed to be Brazil, which became the first developing country outside AOSIS to concede that ultimately the developing countries too might need to assume emissions targets.

The negotiations ended in complete deadlock. There were four months and only five negotiating days to go before the Kyoto summit started.

A strong El Niño had been brewing in the Pacific since April. The temperature rise in the warming waters off Peru was so swift that scientists were soon predicting that this event might be bigger than the 1982–83 one, the El Niño of the Century, which had caused damage around the world totalling $25 billion in today's prices. In June, abnormal trade wind behaviour of a kind not seen since the 1982–83 event appeared in the Pacific. By now, a steady drip of media stories could be seen in the papers. During the summer, financial analysts began to pay attention. Warnings went out to investors to take notice of El Niño when picking stocks in Latin America and Asia. HSBC's economic advisor for emerging markets summarized the situation: 'No one is able to identify the global effects, but the basic picture is that prices will go up, incomes will go down and trade balances will be hit.'

Sure enough, cocoa prices shot through the roof in anticipation that bad weather would add to a global cocoa deficit. In drought-stricken Papua New Guinea, low river levels left a major BHP copper mine unable to handle concentrates, and the company's share price tumbled. These were just the opening shots: the El Niño was not expected to reach its peak until Christmas.

Meanwhile, in the science pages, I read the usual speculation about whether global warming could be strengthening El Niños by making the events more frequent and prolonged. Scientists could not know for sure, the articles tended to say, but there was a risk that this could be the case.

Meanwhile, with Kyoto looming, and despite a steady flow of news coverage about costly weather phenomena, the insurance industry slumbered on. The *Financial Times* ran a survey of the global reinsurance industry on the eve of its annual retreat to Monte Carlo in early September. 'The world's reinsurance industry is living on borrowed time,' it concluded.

The industry rate-cutting competition continued. Why charge high rates, so the predominant peer group sentiment went, when there had not been a sizeable catastrophe in months? An *Insurance Day* editorial chided the industry. 'It seems the minuet will continue, with one side seeking bargains while the other refuses to stand firm for fear of losing competitive ground. One can only hope that in the event of a major loss it will not prove to be a dance to the death for some of the participants.'

*

Sometimes, the scientific work did narrow uncertainties, for example in the case of research on sulphate aerosols. At other times, it seemed that new discoveries only teased us with a snapshot of how ignorant we were. In early September came a most spectacular example of this. Scientists had hitherto assumed that the extent of the Southern Ocean's ice had remained largely unaltered during the twentieth century. But a scientist in Tasmania, Bill de la Mare, now demonstrated conclusively that fully a quarter of the ice had disappeared. His discovery came not from satellites, computer models or ship-borne geophysics, but from whaling records, of all things. These were so complete, and the voyages so sadly numerous, that going back through the years the logs of southernmost catches showed clearly where the front of Antarctica's fringing ice was at any given time. It turned out that an area of ice four times the size of Alaska had been lost over a period of just 15 years, beginning in the late 1950s.

New Scientist magazine went ballistic about the revelation. 'We can't rely on accidental discoveries for vital information about the planet,' raged an editorial. 'Governments can't have it both ways. They whine that they can do nothing about global warming because of "scientific uncertainty", then they cut back on the very science we need to end that uncertainty.'

In the summer of 1997, Greenpeace decided to launch their most difficult campaign yet: an all-out effort to draw a line in the sand over fossil-fuel use. When the world couldn't afford to burn all the oil already found without risking climatic catastrophe, then what was the point in developing new oilfields? That was the question that had to be asked, the group figured. Everyone else seemed nervous of posing the question, yet it was a logical corollary of any effort to stabilize atmospheric greenhouse-gas concentrations in time to guarantee avoiding apocalypse. The way to begin persuading the world to face up to it was to try to stop new oil development on one of the industry's new frontiers.

The battleground chosen was the Atlantic frontier, a deep-water area north-west of Scotland. This area, in which giant oilfields had already been found, was vital to the plans of thirty oil companies. BP was the lead player among them, and hoped to begin production in

October in two fields together containing over 800 million barrels of oil.

Over the summer, Greenpeace and BP edged towards a clash. I watched the battle brewing with a feeling of poignancy. BP had said constructive things about solar, but the bottom line was that they did not want their expanded solar business to progressively replace oil – rather, they intended to increase oil exploration and expand production alongside their ramped-up solar operation. Part of the Greenpeace campaign would be to challenge BP to become serious about solar on a scale that would kick-start the market. They could, for example, build a large manufacturing plant, at a stroke making solar cells cost-competitive with coal. A 100 megawatt per year thin-film plant would do the job. They could build such a market-busting plant for less than the cost of a single leg of a deep-water drilling rig.

On the other hand, Greenpeace had chosen to do battle with the oil company which clearly had the best record of environmental performance. I knew for sure that BP was trying. I also knew that if they did what Greenpeace wanted they would risk taking a catastrophic hit on their share valuation, only to see one of their competitors step in to assume any lost business.

The battle began in mid-August. Four Greenpeace activists occupied a mobile BP rig in the Foinaven field. They sat miserably in a small pod slung below the platform. It was hardly the Brent Spar occupation, but it had the effect of stopping the rig from operating. The Greenpeace PR machine went into action to try to explain why they were doing it. Sure enough, the group was raked by outraged media comment from the outset. The *Financial Times*, for example, found it a 'silly' campaign, based on 'intellectually risible' arguments. This was the tenor of much of the press coverage. The Labour Government, which had been so supportive of Greenpeace in opposition, now advised the oil company on how best to defeat the environmentalists' PR effort: by ignoring it. The *Guardian* came closest to what I considered to be the truth. As one opinion piece in the paper put it, the group had graduated from lucrative campaigns which did not address the planet's survival, to more complicated campaigns 'about things that really matter'.

After eight days the activists were arrested by police. BP then announced it would sue for £1.4 million in damages – the losses it incurred as a result of interrupted business. For a while, I thought they had made the same mistake as Shell: massive overreaction which would have the effect of shining a recurrent spotlight on the campaign, and piling up public support behind the underdog. But within just two days they had backed down.

The campaign quickly subsided from the news pages into the opinion pages. Greenpeace, vowing to continue opposing oil development on the Atlantic frontier, returned to the drawing board to figure out its next effort to focus the public eye on the things that really mattered.

The El Niño marched on. Forest fires were now raging in Indonesia, all across southern Borneo, Irian Jaya and southern Sumatra. The Indonesian authorities had for some time been hiding the scale of the problem. The fires, affecting up to 8,000 square kilometres of rainforest, had mostly been started by plantation owners as a cheap way of clearing land. But the severe drought which had built up by this time as a result of El Niño had sent the fires out of control. Satellite photos showed that the conflagration had spread to deep peatlands under the forest floor, some of them ten metres thick, where a fire could burn for years in the absence of rain. The world's biggest man-made environmental problem was in full swing, now affecting 70 million people in six Southeast Asian countries. In some classrooms, children could not see blackboards. In some cities, cars had to use headlights in the middle of the day. The respiratory toll did not bear thinking about.

Indonesia finally declared a national disaster, and Malaysia a state of emergency. Millions were mobilized to fight the fires, although by now it was clear that only monsoon rains could put them out – but the monsoon was two months late already. There was plenty of rain falling elsewhere in the world. Southern California, for example, was recovering from its first September downpour in fifty years. But there was no rain where it was most needed. And so people died in their hundreds. An airbus crashed, unsighted in the smoke, killing hundreds more. An international team of scientists calculated that if the rains didn't come, and if the peat burned for six months, it would release a billion tonnes of carbon: more than all the cars and power stations of Western Europe emitted in a year.

As the Asian rainforests burned, the pre-Kyoto public-relations battle went into in full swing. In August, an opinion poll commissioned by the Worldwide Fund for Nature held good news. Two-thirds of US voters evidently now believed that global warming was such a threat that they would support an international effort to cut emissions even if it meant hiking energy prices. Fully three-quarters agreed with the statement that 'the only scientists who do not believe global warming is happening are paid by big oil, coal and gas companies'.

The Global Climate Coalition and its allies did their impressive best to counter these views during September, taking out millions of dollars of television advertising across the USA. These adverts attempted to ridicule global warming as an issue, and showed images of developing countries snipped by scissors from a map of the world. This was somehow intended to show that they would be evading their responsibilities, to America's detriment, if the Kyoto treaty went ahead.

At the end of September, 1,500 scientists – including 104 of the 138 living winners of Noble prizes in the sciences – signed a declaration urging world leaders to cut emissions at once.

BP continued its charm offensive. Chief executive John Browne gave another widely reported speech, this time in Germany. 'We see solar as a significant long-term business opportunity,' he said. 'We are testing what is possible by using solar power on a series of our own retail sites.' One, a petrol station in Berlin, would be opened that very day. 'If these tests are successful we will look to extend these solar powered sites across the world.'

This maintained BP's corporate image as the new green industry leader, but at the same time suggested that more needed to be known before solar technology could be used en masse. Half a century after the Manhattan Project, a quarter of a century after the Apollo Project, I wondered how willing people would be to accept that BP needed to run solar panels on a few of their petrol stations before judging whether the technology was ready for wholesale deployment.

Japan, meanwhile, was finally narrowing down on an emissions target. At the end of August an increasingly impatient Prime Minister Ryutaro Hashimoto had instructed his ministries to resolve their differences and agree a target for Kyoto by the end of September. But the deadline came and went. The Environment Agency reportedly wanted a 7 per cent cut in the three main greenhouse gases by 2010, and the Foreign Ministry 6.5 per cent. The Ministry of International Trade and Industry wanted merely to freeze emissions, and would give no ground. Cutting, they said, would seriously harm the Japanese economy. Furious negotiations ensued, and a few days later the compromise was announced: a 5 per cent cut by developed countries between 2008 and 2012.

The European Commission immediately counterattacked. 'This is not nearly enough,' said commission spokesman Peter Jorgensen. Australia did the same, from the other direction. 'These are simply not achievable without enormous costs for Australia,' said Prime Minister John Howard.

The Japanese proposal had a crafty subtext. The 5 per cent total could be trimmed if 1990 emissions per unit of GNP were lower than the developed-country average. This would make the Japanese target closer to 3 per cent. Other exclusion clauses allowed the USA and Australia to be in the same lower bracket.

On 6 October, the White House convened another round-table session on climate change, this time at Georgetown University. 'I think we all have to agree that the potential for serious climate disruption is real,' Bill Clinton said. But he still would not reveal the US target, and he warned that the developing countries would have to play their part in the solution. In Kyoto, the USA would be asking for 'meaningful but equitable commitments from all nations'. What he meant by that was entirely unclear.

The US Department of Energy had recently reported that the United States could cut carbon emissions to 1990 levels by 2010 with no net loss to the economy. The way to do that was to invest in two hundred energy-efficiency technologies spread across the building, transportation, industry and utility sectors. It would cost $50–90 billion to develop and switch to the technologies, and that would be around the sum of the savings on the national energy bill. This was somewhat different from the US Chamber of Commerce view, which estimated the price tag to be $277 billion, or the American Petroleum Institute estimate, which put it as high as $350 billion per year. The Chamber's president, Thomas Donohue, warned that 'America's jobs and economy are being held hostage to the environmental-political agenda of the United Nations.' The administration was perpetuating a 'myth that "free lunch" technologies' were available, said the API's William O'Keefe, who now doubled as chairman of the Global Climate Coalition.

CNN decided they had had enough of this kind of nonsense. They pulled all TV adverts on climate change.

In early October, the Food and Agriculture Organization warned that the El Niño was now giving cause for concern over food supplies in many Asian and Pacific Rim countries. NASA deemed the event the worst this century. States of emergency had been declared in countries on both sides of the Pacific over the drought threat to crops. The indirect effects were just as appalling. Fed by the worst drought in half a century, 10,000 fires raging across Indonesia showed on satellite images, with new hot spots appearing every day.

Forecasts of damage from the El Niño now topped $20 billion in flood and drought damage. Storm after storm was being spawned in the Pacific, and September saw Linda, the strongest cyclone ever recorded. For a while, this monster threatened California, where a hurricane had never made land. Yet in the Atlantic it was the quietest hurricane season in years, as would be expected in a strong El Niño year. 'My response is Hallelujah! My best friend is El Niño,' Florida Insurance Commissioner Bill Nelson told Reuters. But at the same time he testified to Florida's Senate Banking and Insurance Committee that there was no overall decrease in risk, and indeed that the risk merited a large increase in the state's hurricane trust fund. Otherwise, Nelson said, come 1999, when the moratorium on insurance cancellations ended, there would be a mass exodus of insurers from the state.

Bermudan reinsurer Mid Ocean Re, for one, didn't seem too concerned. 'Despite all the storm activity and drought, it [the El Niño] hasn't been a major insured, or reinsured, event,' a spokesman said. And with the end of the year rapidly approaching, the US insurance industry looked to be heading for bumper profits. Allstate's third-quarter net income, for example, was $825 million, up from $292 in the third quarter of 1996 – itself not such a bad year.

In this environment, it now seemed less and less likely that the insurance industry would become a substantive force for progress at Kyoto. *Insurance Day* regaled its readers once again. 'For the sake of their own businesses,' raged an editorial on climate change, 'insurers must be prepared to deliver by lobbying long, hard and visibly.'

But I already knew that the most I could hope for in Kyoto was another short flying visit by a handful of insurers.

It now seemed increasingly likely that the most effective voice for progress in Kyoto might actually come from within the oil industry. On the same day the White House held its seminar, Shell announced it would be investing $250 million in renewable energy over the next five years. The previous week it had opened a solar-cell production line in the Netherlands. It seemed that Shell was in the process of emulating BP. Indeed, it was even possible that they were trying to outdo their rival. The aim, so a company press release said, was to make renewable energy a 'fifth core business', alongside oil exploration, oil production, chemicals, gas and coal. A new business, Shell International Renewables, was in the process of being set up.

Ten days later, Shell gave more details. In the interim, the amount to be invested had increased to $500 million. The president of the new renewables company, Jim Dawson, said that sales of renewables would be almost $250 billion dollars by 2020, up from $10 billion today. Shell's target was to have 10 per cent of the market by 2005. The investment was to be split evenly between solar and biomass.

This initial investment was still very modest compared with the $10 billion invested annually in all the other Shell Group activities, but it was clearly a development of the most profound geopolitical significance. The opposition was now well and truly split, with the European oil majors lending credibility to the environmentalists' cause with virtually everything they said. As Jim Dawson put it, this was real – Shell was undergoing a 'step change'.

But just as BP and Shell stepped firmly towards the environmental side, so the American oil companies dug in deeper. Exxon Chief

Executive Lee Raymond, addressing the World Petroleum Congress in Beijing in mid-October, gave the most irresponsible speech yet by an oil chief. He rubbished global warming using the crudest of the disgraced dissident arguments, and called attempts to curtail fossil-fuel use 'neither prudent nor practical'. He specifically included in that assessment Chinese coal. The use of fossil fuels was essential, he said, for both economic growth and the eradication of poverty. As for oil, China should maintain and if possible increase its local production.

This was quite blatant hypocrisy. While in Beijing Raymond was exhorting China to burn away, in Washington Exxon was financing the Global Climate Coalition's key effort to wreck Kyoto by insisting that the developing countries take on emissions reductions commitments of their own.

But Exxon was beginning to look dangerously out of tune. By now even the automobile giants were having second thoughts, it seemed. William Clay Ford, Jr, an executive at Ford Motor Company, gave a speech on 16 October saying that global warming was a genuine threat to the environment and that those automakers who said otherwise were in danger of being 'marginalized in the court of public opinion'. This was all the more significant for the fact that William Ford was the great-grandson of Henry Ford himself, and he aspired to become the next chairman of his ancestor's company.

On 21 October, General Motors joined the list of those seemingly flirting with defection from the carbon club. Said chairman Jack Smith: 'We recognize that there has been an increase in CO_2. It is cause for concern. And we feel strongly that there needs to be a significant effort to improve the technology that will reduce CO_2 emissions.' GM, he said, would be introducing some of these at an international auto show in Detroit in January.

These developments were encouraging, but not to be overstated. Both Ford and Smith had trotted out the mantra about the developing countries having to act alongside the USA. This remained the issue capable of wrecking the Kyoto protocol, and hence the Convention on Climate Change, from within. It was the trump card that the carbon club would play in the last session of negotiations in Bonn, and at Kyoto itself. And of course, Ford and GM would still be there as members of the Global Climate Coalition alongside Exxon, Mobil and the rest.

OCTOBER 1997, BONN

The tide of opinion really did seem to be turning in the United States. In the first few weeks of October, editorial boards across the nation issued calls for action. 'Stronger leadership is warranted,' concluded the *Miami Herald*. 'Far more dramatic steps are needed to turn down the heat,' said the *Minneapolis Star Tribune*. 'Good intentions are not enough,' opined the *San Francisco Chronicle*. The *Boston Globe* was interested in the reasons why US leadership had not already been in evidence: 'the Government needs to resist forces in industry'.

Climate negotiators from almost every country in the world gathered in Bonn on 17 October. They faced a week of preliminary discussions, and then a week-long formal session of the Ad-hoc Group on the Berlin Mandate, the eighth of its kind. On the Wednesday of the first week, the developing countries presented their combined proposal. The Kyoto Protocol should require these so-called Annex 1 countries – the countries listed in Annex 1 of the Convention on Climate Change – to take on 15 per cent cuts in carbon dioxide, methane and nitrous oxide, from 1990 levels, by 2010, and 35 per cent by 2020. This was the bare minimum that would be needed to persuade the developing world that the developed world was taking climate change seriously. Nearly 150 countries stood behind this proposal. The G77 and China, OPEC and AOSIS, seemed finally to have identified common ground. As for their own contribution, any talking about their emissions would have to wait until later, as the Berlin Mandate specified, and after the developed world had genuinely begun the process of cutting its own emissions.

The same day, the USA announced its own proposal for a Kyoto Protocol. The Clinton Administration asked the Annex 1 countries merely to freeze greenhouse-gas emissions at 1990 levels averaged over a budget period from 2008 to 2010. If they could agree this, then the USA would sign a Kyoto Protocol – provided, that is, the developing countries could themselves agree to 'meaningful participation'.

It seemed that the president had decided he wanted to put off cuts for twenty years from the time that most industrialized nations first pledged them. Under this proposal, the first cuts in greenhouse gases would come not even in the next term to his, but three presidential terms hence. Moreover, overall emissions in Annex 1 countries were already some 5 per cent below 1990 levels, meaning that even an emissions stabilization target would actually be a licence to increase.

The main sop in the weak White House position was the offer of $5 billion over five years in incentives for voluntary action, much of it as tax breaks for the use of low-emissions technology. This sum, the president said, would be included in the 1999 budget: $1 billion a year in a $5,000 billion per annum economy, for voluntary action, aiming only at freezing emissions.

The day immediately became known in Bonn as Black Wednesday. The reaction from environmental NGOs was one of uniform rage. The daily newsletter *Eco* fumed at the White House, conveying vitriolic rhetoric from non-American and American environmentalists alike. The Clinton negotiators were radiating an air of 'take it or leave it', just as had been the case in the worst of the Bush years. If the US target was a line in the sand beyond which it was not prepared to go, an *Eco* editorial said, it should just say so. That way the world would be spared all the greenhouse-gas emissions of the trip to Kyoto, and the Japanese government would be spared the political anguish of playing host to a fiasco.

Leading European Union negotiators professed themselves sick with disappointment that the USA was being so backward. On the Friday of the first week, the European Commission gave a seminar on how the EU could meet its target, and do so at a maximum cost of 0.4 per cent of GDP. The Americans, having said for so long that the European target was unrealistic, didn't even bother to send anyone to the briefing to check whether it was indeed so.

Japanese Prime Minister Hashimoto seemed to agree about the inadequacy of the target. 'I think there might have been room for further efforts,' he said.

On the Saturday, the American delegation walked out of a contact group on trading and joint implementation, ostensibly because of the absence of certain key delegates from China and India. They left fury in their wake. When I arrived in Bonn on the Sunday, former colleagues and friends on the delegations told me that the atmosphere at the talks was as bad as it ever had been. The overall effect of the Americans' intransigence had been simply to gum up the whole process, with nations retreating into old positions. The climate talks were no longer, effectively, negotiations. AGBM8 was doomed to be a forum for trading hostile rhetoric.

As the preparatory week in Bonn drew to a close, further black clouds were gathering on the horizon, even if a decent emissions-reduction target could ever be agreed. Clear potential loopholes were under consideration by negotiators. The first of these involved the three fluorine-bearing greenhouse gases, hydrofluorocarbons (HFCs), perfluorocarbons (PFCs), and sulphur hexafluoride (SF_6). Let us call these, as environmentalists tend to, the 'F' gases. When added to the EU's total, by one estimate of uses of the F gases, the 15 per cent overall target for the three main greenhouse gases (CO_2, nitrous oxide and methane) would shrink to 5 per cent below the 1990 baseline, and Japan's 2.6 per cent emissions reduction proposal would be transformed into to a 7 per cent rise.

The second potential loophole involved so-called 'hot air' trading with Russia. Because it was likely that by 2010 the Russians would still be emitting at less than their 1990 baseline, they would have nominal emissions reductions to trade with the West. If the USA opted to buy such emissions trades, it would in principle actually be able to increase emissions by 5 per cent and still register a zero increase under a protocol, having done nothing at home.

Third came the net approach. New Zealand had long been an advocate of allowing nations to count carbon sequestered in vegetation as credit against carbon dioxide reductions. The USA, it seemed, was now viewing favourably a refinement of the New Zealand special plea known as 'gross-net'. In this scheme, countries would calculate their emissions beginning with the gross total for 1990 (not including sinks) and then deduct their national sink storage of carbon from the emissions budget period. Adopting this method would result in a 7 per cent increase in US emissions.

By exploiting all three loopholes, the overall real US *increase* in emissions might therefore be around 12 per cent, with the government still legally entitled to claim that it was achieving a net freeze.

On top of all this would come the exclusion of emissions from international aviation and marine bunker fuels, which most governments felt at this stage were too complicated, in terms of assigning ownership, to include in any protocol. But their contribution to global emissions would inevitably be large: by 2010, their growth alone could be in the 5–10 per cent range for the developed countries.

Finally, on top of all this came the dreadful prospect of emissions borrowing. Under this proposal, nations which failed to meet their emissions target would be allowed to borrow from future emissions allowances without penalty. This proposal offered the spectre of another toothless treaty, with a catch-all get-out clause that would remove all urgency from policymaking.

I turned up for AGBM8 with a heavy heart. Mingling in the foyer of the negotiating hall, I came across several of my old carbon-club antagonists in quick succession. J. R. Spradley scrutinized my name badge, which showed me to be a representative of The Solar Century. 'So you're getting sun in the UK these days,' he said. I asked him if the recent solar advocacy by both BP and Shell wasn't shaking his assurance in coal a little. It wasn't, apparently.

Mobil's Leonard Bernstein greeted me with a sneer, and I asked him if he was pleased with his progress. He nodded. 'Some of the economic realities are dawning,' he said.

I came upon Exxon's Brian Flannery engaged in a disagreement with Mac MacFarlane of DuPont. Flannery nodded at me as he walked off. I asked MacFarlane what the altercation had been about. MacFarlane smirked. 'Let's just call it a difference of philosophy,' he said.

The DuPont man was someone I had watched evolving his views over the years as the evidence mounted. In fact, he now worked on secondment for UNEP, as a highly valued policy advisor on ozone depletion and climate change. There was little he and Flannery could agree on any more.

Soon I came upon another man who had switched position. BP's

Klaus Kohlhaus was now firmly in a different camp. 'Things are moving,' he told me, with an air approaching enthusiasm. 'It's possible to talk about this issue now.' He meant in his own company.

But people like Kohlhaus and MacFarlane were still rare among the industry NGOs. The carbon club was fielding no fewer than 80 lobbyists in Bonn. And the devil of it all was that they seemed, given President Clinton's latest studied piece of fence-sitting, to be winning. What Mobil's Bernstein meant by 'economic realities dawning' was that he thought the prospects of a protocol were dying. Others agreed. Before AGBM8 had finished, the first report appeared in a national newspaper that the Kyoto Protocol was dead. 'The corporate lobby on three continents has systematically tried to sabotage any cuts in greenhouse-gas emissions,' the UK *Guardian*'s environment team wrote. 'Last week it triumphed. The champagne corks are popping in the boardrooms of BP, Shell, Esso, Mobil, Ford, General Motors and the coal, steel and aluminium corporations of the US, Australia and Europe.'

On the Tuesday, Chairman Raul Estrada gave a sober briefing to more than a hundred NGOs, the vast majority of them from industry. The previous night's session, he said 'was like time travel. I encountered attitudes I last heard 30 years ago when I began in the multilateral business.' The American retrenchment had caused an eruption of all the old resentments harboured by the South about the North.

On the Wednesday, the Business Council for a Sustainable Energy Future defiantly staked out its case. 'Yes to Kyoto,' was the message, and there to support it were senior executives from Enron, Pilkington Solar and a new player – no less a corporate giant than Daimler Benz. The German car-maker's representative waxed lyrical about the scope for fuel cells to reduce emissions in the transport sector, how exciting the technical developments were in the field, and how his company planned mass production of fuel-cell cars by 2004. Meanwhile, thirteen business leaders gamely took out an advert in US papers challenging their business colleagues to be more constructive about climate change. 'Left uncontrolled, a changing climate could threaten the future prosperity of our nation and the world,' it read. 'To reduce this threat we call upon the United States government to provide strong leadership by promoting climate change policies that provide incentives to act quickly.' The signatories in this minority included Bechtel, Nike, Interface and Mitsubishi Motor Sales of America.

The week dragged on in stalemate to its painful and predictable conclusion. In the final plenary session, the AOSIS chairman, Ambassador Slade of Western Samoa, spoke from the heart. 'Amidst all the perplexities, let us not lose sight of the moral dimension. Let not the many be sacrificed for the few,' he said. 'There would be moral repugnance in that.'

Tanzania spoke on behalf of the G77 and China. The developing countries had overcome significant differences, their ambassador said. Why not the industrialized countries? 'The 79 per cent of countries that will struggle to survive demand action now.'

Zimbabwe, speaking for Africa, warned of agricultural systems in danger of collapse.

The final negotiating text was a ghastly mess, a mass of bracketed language with every major issue still unresolved. The most extreme of the loophole issues – superheated air – was still alive, raised in the last session by Poland, who wanted to bank their emissions reductions made before the budget period and then trade them in the budget period. Russia – unsurprisingly, and Canada – incredibly, offered support to this. If this selfish group got their way, there would be no chance of global emissions reductions.

An eleventh-hour hope that the USA would pull back from the brink came and went. The head of the US delegation, Jonathon Pershing, made a series of mobile phone calls amid much whispering about what this might mean. Then he intervened to insist that text be inserted on the need for developing countries to take on commitments. Canada and New Zealand supported the USA at once.

And so Bonn ended with negotiators in deadlock, facing the prospect of a fiasco in Kyoto, and, even if they could agree a protocol, in grave danger of producing one that sanctioned increases in emissions.

Outside, as the exhausted and mostly grim-faced negotiators dispersed, I watched, sickened, as Don Pearlman shamelessly pumped hands with the Saudi and Kuwaiti delegates.

*

As I watched the carbon club at work that week, it struck me that an entirely new form of conflict had broken out in the world. It threw up all the divisions and emotions of civil war. Governments were divided within. Business groupings were polarized. The hydrocarbon sector, in particular, plainly considered itself to be in a state of civil war. But the divisions were also global. Whole blocs, regions and economy types were lined up against one another. European Union progressives locked horns with historical allies in North America and Australasia. Within the developing world bloc, OPEC lined up against fellow-members with no oil. Island states pointed accusingly at the industrialized world and talked of cultural genocide. The diplomatic rituals of the climate talks were now barely veiling the strongest of feelings.

Adding to the complexity of this drama, players could and did switch sides. Australia had once stood in the trenches with the progressives. Now her diplomats were working with the most merciless of the carbon club's snipers. But through this multi-dimensional battleground, complex as it was and is geopolitically, in essence two sides could be detected in the run-up to Kyoto: those who thought that global warming was a problem that humankind could no longer afford to ignore; and those who didn't think so or, for some unfathomable reason, didn't seem to care.

Is the metaphor of war – the carbon war – an overstatement? How can it be? After all, though we trust no actual shots will be fired, the UN is telling us that the casualties will be measured in hundreds of millions in the decades ahead if greenhouse-gas emissions are not cut significantly. And like many a war, there can be no winners if hostilities persist. The carbon club maintains that it is defending business and national interests. But according to the best available threat assessments, the consequences of the angry tide of climate change will in time wash over economies, grabbing territory and laying waste with the ferocity of the most efficient invading army. Nobody can do good business in the midst of ruins, not even Exxon.

How then, in this context, are we to view the carbon club's black efforts to derail the negotiations – extending as they do well beyond the time when reasonable doubt about the enormity of the risks has evaporated? How else, with stakes such as these, and with the writing so clear on the wall, but as a new form of crime against humanity?

OCTOBER 1997, CITY OF LONDON

During October, I fought the nagging pull of a despair rooted in the feeling that despite all the evidence, and all the talk, governments would not prove capable of responding to the threat of global warming: indeed, of even taking the first tentative step in responding. And this despite the fact that the climate of opinion finally seemed to have turned. *Business Week*, on 3 November, ran an article entitled 'Global warming: Is there still any room for doubt?' Even this most conservative of business organs did not see any. Among the eruption of news coverage in the weeks before the summit were other remarkable indicators of change in the air. A special pre-Kyoto environmental issue of *Time* magazine was laced with advertisements by Toyota for low- and zero-emission cars. Company President Hiroshi Okuda explained why in one of the adverts. 'Consumers are smart. They recognize the threat that pollution and global warming present to them and their children.' Toyota was in the process of doing something about it, it seemed. Other advertisements told us that Toyota was on the verge of bringing to market a hybrid battery/gasoline car, cutting carbon dioxide emissions by 50 per cent, and had recently introduced the world's first commercially produced automobile running on nickel metal hydride batteries. Of course, the extent to which battery cars would deliver depended on how electricity would be generated, the advert continued. Pictures of solar panels and windfarms elsewhere in the issue showed how *Time*'s publishers evidently hoped that would be done.

I busied myself working on a British industry contribution to the debate, with a specific solar-energy spin. Before the Labour Party had come to power, the future ministers of environment and industry in the UK, Michael Meacher and John Battle, had asked me to convene an ad hoc group of companies to advise them on building a British PV market and industry. The idea was that it would not be simply a choir of known PV advocates, but that there should be one company from each of a range of industry sectors: a bank, an insurer, a utility, an architect, an engineer, and so on. It was to be an Industry Solar Taskforce, as opposed to a Solar Industry Taskforce. Over the spring and summer I had managed to assemble an array of key names. BP Solar, Eastern Electricity, Foster and Partners, General Accident, Guardian Royal Exchange and NatWest were among those who agreed to serve.

We submitted a statement of advice to the government on 30 October 1997, aiming to leave them enough time to digest it before Kyoto. I was thrilled with the wording this diverse group managed to agree to. 'Among a wide range of sectors threatened,' the statement said of global warming, 'the insurance and banking sectors face serious problems, and because London is the world's major financial centre, unmitigated enhancement of the greenhouse effect is a threat to the heart of the British economy.'

So what to do about it? 'We recognize the full family of renewable and efficient-energy technologies as being vital in the abatement of this risk, but one technology should be particularly important in the energy mix of a sustainable future: solar photovoltaic. Solar PV could be the single most important long-term means of achieving the deep cuts in greenhouse-gas emissions which are the ultimate agreed objective of the Convention on Climate Change.'

It looked at this time as though the global sales of PV were going to exceed 100 MW for the first time in 1997. (They did. Indeed, they exceeded 120 MW, a market growth of 40 per cent on 1996.) But still, this market had an annual sales volume equivalent to less than a tenth of the electricity generated in single average coal-fired power station. So how could the companies on my taskforce conclude that PV would be so important in the future? The answer lay primarily in the markets of the developing world. A key global imperative, if governments are ever to reach the agreed objective of the Convention on Climate Change, will be the delivery of sustainable energy for development in the developing world, where 2 billion people currently have no electricity, and where those who have it will always need more, however much they improve their efficiency of use. The most realistic form of alternative supply, especially in rural settings away from current electricity grids, is solar PV. Hence, stating the taskforce case another way, we need to fashion huge global solar PV markets in order to win the endgame in the battle against global warming.

In early November, as though on cue, came the announcement of plans to build the world's largest solar PV manufacturing plant. Shell would be joining with Pilkington Solar to build the 25 MW factory in Gelsenkirchen, Germany, providing high-tech jobs for over 300 people. This was still a small plant, compared with what was possible in principle, but it was another step in the right direction.

In mid-November came further support for action from European business. The UK's Confederation of British Industry released the results of a poll of its members on Kyoto. Fully 83 per cent of the member companies expressed support for the EU's 15 per cent target, and 62 per cent believed that Europe should pursue the goal unilaterally if the other industrialized nations refused to adopt it.

On the same day, Shell UK's chief executive, Chris Fay, joined the progressives. Speaking at a conference in London, he said it was time to stop the 'risky experiment' of pumping carbon dioxide into the atmosphere. 'Shell believes that the time for precautionary actions to prevent possible climate change has come.' Fay went a step further than BP by saying that the company now thought that a carbon tax might be a good weapon with which to attack climate change. A week later Shell upped the ante still more, when Group Chairman Cor Herkstroter announced that the group was calling for a global target of 5 per cent for reducing emissions of carbon dioxide. As evidence of its seriousness of intent, Herkstroter said that the group was considering disposing of its coal assets.

But encouraging as this green-image competition between Shell and BP was, one could still see the other side of the carbon coin. Shell produced 12 million tonnes of coal a year, which generated less than 1 per cent of its $8-billion-plus earnings. Meanwhile, in investment-starved Russia, Boris Yeltsin had been forced to sign a decree allowing foreigners to buy Russian oil companies outright. By opening up its oil-and-gas sector in this way, the government hoped to raise $20 billion within a year. Sure enough, within days Shell and BP had concluded deals. Shell had teamed up with Gazprom, the giant gas monopoly, to bid for Rosneft, the last big state oil group to be privatized. BP had bought 10 per cent of another Russian oil company, Sidanco.

*

With just a week to go before the negotiations, the man everyone assumed would head the US delegation, Tim Wirth, resigned. Undersecretary of State Wirth, a long-term advocate of action on global warming, had presided over US delegations at both previous annual Conferences of the Parties. He had been at the helm when the USA gave eleventh-hour ground on both occasions, so enabling the Berlin Mandate and the Geneva Declaration to be agreed. Now he was out. Any way you looked at it, this was discouraging news. The official named to replace him was Undersecretary of State Stuart Eizenstat, a veteran of international trade and economic negotiations but with no known interest in the environment. Talking to the press the Thursday before the Summit began, Eizenstat was hawkish. 'We are not going there for an agreement at any cost,' he emphasized. A deal would have to have reasonable targets, including emissions trading and net accounting. The USA still had 'very serious problems' with the EU bubble. And of course, there would have to be meaningful participation from developing countries. Otherwise a US walkout was not out of the question.

Rumour had it in Washington that Al Gore was now not going to Kyoto, in order to keep clear of what now seemed set to be a fiasco.

11

The Day of the Atmosphere

November–December 1997

KYOTO, JAPAN

In the wooded hills fringing the ancient city, autumn was lingering to welcome the thousands attending the Kyoto Climate Summit. Centuries-old temples vied for wintry sunlight in the roofscape as I looked across the city from my hotel to the wooded ridges. You could see at a glance why the Japanese regard Kyoto as such a special place. I made a wish that it would become an even more special place. That within the next ten days the rusty hues unseasonably splashing the hills would become an emblem for the beginning of another autumn – that of oil and coal.

For the carbon club, as for the environmentalists, this was a defining battle. If the carbon club won here, they stood to knock the whole process off the rails. If they lost, and a protocol – any protocol – with legally binding cuts was negotiated, they would probably always be on the defensive thereafter: defending one trench after another as they were pushed further and further back. Almost all the central players from the pre-Kyoto phase of the carbon war were in town. The new executive director of the Global Climate Coalition, Gail McDonald, presided over a delegation of no fewer than 63, including representatives from the American Petroleum Institute, the American Automobile Manufacturers Association and the US National Coal Association. Don Pearlman led his Climate Council band. The main American oil companies had their usual representatives. Brian Flannery was there for Exxon to play his scientific sceptic role, as was Fred Singer, batting for all fossil-fuel comers. Leonard Bernstein was in town to push Mobil's propaganda. At least his current crop had the benefit of clarity. 'We oppose legally binding targets and timetables at this time,' the company's position paper stated. 'We don't believe we should rush to a potentially damaging solution based on an uncertain premise.' What a contrast to BP's Kyoto position statement: 'The prospect of global climate change is a matter for genuine public concern. We share this concern . . . This debate is too important for us to stand on the sidelines or just say "no".'

Heading the International Chamber of Commerce delegation of over a hundred, as he had in Berlin, was Texaco's Clem Malin. My best efforts after Berlin to point out to the ICC the potential conflict of interest in Texaco's public relations chief speaking for 7,500 businesses and associations in 130 countries had failed. He was still there, and his fingerprints were all over the ICC position statement. This may not have had the shrill tone of a Global Climate Coalition tract, but it subtly pushed voluntary actions as the best way to go. All forms of energy would be required *throughout* the twenty-first century. And of course, developing countries should accept the need for their participation.

Aided on the ground by particularly vocal supporters in the US Senate, led by Senator Chuck Hagel, the carbon club would be doing all it could to defend the wrecking position it had built up around participation by the developing countries.

The Kyoto international conference centre was a squat concrete complex amid leafy hills and carp-filled lakes in the north of the city. Registrations on the first morning included 1,500 government delegates from 160 countries and 3,500 journalists from over 400 media organizations. Only 600 seats were available for the latter in the main hall. Meanwhile, 3,600 observers – environmental and business NGOs plus intergovernmental agencies – would be vying for fewer than 400 seats. To ensure that this did not result in mayhem, the UN had installed large TV screens at various points around the conference complex. These screens would show live coverage of the negotiating sessions, plus press conferences. They would become the focal points for much of the drama in the days that followed. The media and NGOs had been mixed in together in a hall the size of a small football stadium. One would grow used to seeing crowded press conferences in the main press room, yet hundreds more journalists of all nationalities crowded round TV screens listening in the working hall. They would then return to their positions among the rows of long working tables, where they would endeavour to work on their laptops, each amid an untidy sprawl of press releases from the NGOs. As the reporters sat there, feverishly typing or frowningly awaiting inspiration, yet more press releases would be added to their individual heaps. Mobile phones trilled incessantly and the hundreds of conversations merged into a constant hum. From the balcony, the hall looked and sounded like a heaving human equivalent of an anthill.

Only in Rio at the Earth Summit had I seen anything quite like it. This time, however, the negotiations took place on an IT revolution wave that had been but a ripple in 1992. The interest in Kyoto around the world was evidently huge, and it was being met by broadcasts of the open sessions live on the Internet. Scheduled interactive chat sessions allowed remote interaction with experts and other guests of the UN in Kyoto.

I watched the Greenpeace delegation at work, a 45-strong party from ten countries. I didn't miss the stress my friends and former colleagues were under, but I did miss the companionship, the sense of being part of a team. I would not be in their meetings, or even in the environmental NGO coordination meetings. I would be attending the business NGO meetings.

That first morning, as I renewed my acquaintance with a friend from AOSIS, I felt a tap on my shoulder. I turned round to see Don Pearlman standing there, an expression of what he seemed to hope was playful greeting on his face. His hand was extended. I shook it by reflex, but regretted doing so at once.

He looked down at my nametag. 'And what are you these days?'

I hadn't spoken to the high priest of the carbon club at any length since Berlin, two years before. I had resolved to end the jocular interactions of the past years when I had read about his credentials in *Der Spiegel*. His role in the vile attack on Ben Santer in 1996 had cemented my view.

I looked over his shoulder, weighing the possible responses. Dammit, this was war. He had said so himself. This was an amoral man doing evil work. Why pretend otherwise?

'I really don't think I want to be talking to you, Don.'

'OK, I wouldn't want to embarrass you.' The jowls descended, and he departed.

Day One, Monday 1 December, began with the normal speechmaking from the leading lights. Ambassador Raul Estrada, Argentinian master diplomat, was the man with the awesome responsibility of trying to steer this summit to success. He would chair the hard-nosed negotiating sessions, referred to as the Conference of the Whole, while the Chairman for the plenaries would be Japan's Environment Agency chief, Hiroshi Oki. In his welcoming speech, Oki called for a spirit of cooperation and 'friendly concessions'.

The US delegation was led for the moment by Melinda Kimble, an acting assistant secretary of state, and not even a Clinton appointee. Not only had the recently appointed Assistant Secretary Eizenstat yet to arrive, but it was still not clear whether Vice-President Gore would be attending. I found this incredible. How could he stay away?

Kimble gave a bullish speech insisting that all six gases (CO_2, methane, nitrous oxide, hydrofluorocarbons, perfluorocarbons and sulphur hexafluoride) plus sinks be included in the protocol, along with trading and joint implementation for credit. She cast the US target as a decrease in US emissions of approximately 30 per cent below levels we would expect to see in 2010 in the absence of a protocol. This, she said – echoing a mantra of US spokespersons in the preceding weeks – was a significant effort, an effort comparable to that offered by any other party.

It was disingenuousness of the saddest kind. First, these were not real emissions reductions. They were reductions in Department of Energy projections for emissions growth – guesstimates twelve years into the future, in essence. Moreover, with all the loopholes the US advocated, how could they know what their effort would come to by 2010? How did they know that US emissions wouldn't actually rise by 2010, if they did enough trading, joint implementation and offsetting of emissions against forestry? Emissions in Germany, the UK, Denmark and the rest of the EU would certainly not be rising if the bubble target was to be hit.

But Kimble attacked the bubble too, listing five areas of 'strong concern'.

There was one sign of potential compromise. Although the US still advocated a flat-rate target, it was prepared to consider the possibility of limited, carefully bounded differentiation of country targets, Kimble said. However, neither in the speech nor in the ensuing press briefing did she elaborate on that.

Luxembourg, speaking in its role as president of the EU, was brief and to the point. 'After two years of negotiations,' Ambassador Pierre Gremagna began, 'we may have lost sight of the reason we are here.' We must, he emphasized, *strengthen* the commitment made in 1992. In a press briefing after the session, Gremagna was forthright. The new US flexibility was in the wrong direction, he said. 'We get the impression that the game is to find ever more loopholes in these negotiations, and that is a bad omen. We need credible targets.'

On Day Two, another player looking at multibillion-dollar stakes began its contribution to the Kyoto process. The global insurance industry had taken $2,000 billion in premium income over the previous year. Now over 200 insurance and banking executives from 10 countries gathered in Tokyo to consider the implications of climate change for that considerable pot of capital, hoping to send a clear message to Kyoto. General Accident, Gerling, National Provident Institution, Storebrand, Sumitomo and Swiss Re would be sending representatives down to Kyoto later in the week to relay the results.

'One storm could do $100 billion of damage in 100 hours, roughly half of it insured,' said General Accident's Andrew Dlugolecki, opening the proceedings as usual. 'People still haven't grasped this.'

As the day proceeded, it became clear that people within the threatened industry itself still didn't seem to grasp the reality. The two Japanese companies prominent in the initiative, Sumitomo and Yasuda, seemed much less committed to action than their European counterparts. At their insistence, language in the insurers' declaration for Kyoto which in draft form called for urgent agreement of substantial emissions reductions was watered down to the following weak passage: 'measures which will decouple the emission pathway from a business-as-usual scenario.' Several of my European insurer friends were seething with frustration at this development.

The sad truth was that the insurance industry, for all my hopes – and the promise of events in 1995 – had not evolved into a force capable of exerting any serious pressure on the Kyoto process. After five years of work, I had to face the fact that the serious will to act within the financial services sector was still limited to a depressingly small cadre of well-informed individuals. The insurers behind the UNEP initiative had tried and failed to persuade their boardrooms to back an effective proactive resolution. For all the promising rhetoric, the insurance industry had yet to agree to field a single full-time representative of their interests against the dozens from the oil, coal and automobile industries at the climate talks. For a $2,000 billion industry – bigger than coal and oil combined – this would have been laughable if it wasn't such a tragic lost opportunity.

Elizabeth Dowdeswell, UNEP's executive director, did what she could to coax the sector along. 'I want to challenge your companies,' she said in her opening speech. 'When the time is right, we would like you to join a cross-sectoral business alliance for progress on climate change.'

But the time *was* right, and had long since been so. That such a cross-sectoral alliance was possible in practice, the UK Industry Solar Taskforce had already perhaps shown in miniature. All that was needed was an international version of the British taskforce covering the full gamut of clean-energy technologies.

And it was not as though erosion of profitability was all that was at stake. The keynote speaker at the UNEP event was the ex-director general of global environment at the Japanese Environment Agency, Saburo Kato. Since leaving government service for the world of think-tanks, Kato had made space for himself to speak out, and today he did just that. 'It seems clear that environmental problems have the potential to irreversibly destroy the conditions necessary to support not only human life but all life on earth in the twenty-first century,' he told the assembled insurers. This, he argued, offered a little added value to the business case for proactive action.

*

At this stage both New Zealand and Canada announced their targets: 5 and 3 per cent reductions by 2010, respectively, including a permissive array of loopholes. It was beginning to look like the JUSCANZ group was zeroing in on a 3 per cent target. Japan had long since tabled a 3 per cent differentiated target for itself and 5 per cent for most other Annex 1 countries. The USA could easily move from zero to 3 per cent to stay in line with their group. Yet Annex 1 countries were today already collectively 4.6 per cent below 1990 emissions levels, largely as a result of eastern European economic decline. A 3 per cent target by 2010 would therefore represent a *rise* in emissions, notwithstanding loopholes.

More bad news came from the discussion group on targets. The USA and Canada had offered little resistance to a suggestion that the protocol embrace only carbon dioxide, methane and nitrous oxide in 1997, and leave consideration of the three 'F' gases until 1998 or later. It seemed that the USA might be about to negotiate away the only strong suit it held relative to the EU.

What good news the day had to offer came from the cities. The International Council for Local Environmental Initiatives gave a press briefing, calling on governments to adopt a 20 per cent target by 2010 like many of their members were going to do. They had 200 municipalities from 29 countries organized in their Cities for Climate Protection campaign – over 100 million people, representing fully 5 per cent of global greenhouse-gas emissions. The local governments would be pressing ahead whatever the national governments did.

When I arrived at the negotiations on Day Three, I learned the good news that Vice-President Gore would indeed be coming to Kyoto. The bad news was that he would come for one day only, the Monday of week two, and that he seemed to be defusing expectations. 'I would make it clear,' he told a press briefing in the White House Cabinet Room, 'that, as others have said, we are perfectly prepared to walk away from an agreement that we don't think will work.'

Writing in the *Washington Post*, the feisty former British environment minister, John Gummer, sounded the alarm from across the Atlantic. 'The entire scheme seems inspired by a misplaced optimism that "something will turn up", as if global warming was a bad dream.'

The Conservative member of parliament had turned into a bitter critic of conservative America in recent months, and now he lambasted American industry. 'In a peculiar way, American business is behaving like old-fashioned socialists by trying to protect itself from innovation.'

The *New York Times* had conducted a large opinion survey on global warming the previous week. Its results were offering encouragement to the US environment groups in Kyoto. Fully 65 per cent of those polled felt that the USA should take steps to cut its own emissions immediately, irrespective of what others did. The margin of error on this figure was just 3 per cent. It seemed, as the *Times* put it, that 'the American people are far more willing than their government to take early, unilateral, steps'. As for the public's response to oil- and coal-industry arguments that emissions reductions would be economically ruinous, 'they appear to be unimpressed', the *Times* concluded.

If I had just spent $13 million on an advertising campaign pushing the reductions-ruin-economies line, as US industry groups had, I would have hated to read that.

Meanwhile, progressives in US industry were building a steady backlash. The pro-Kyoto statement signed by business leaders that had appeared in Bonn now had more than sixty signatories. This statement, exhorting the Clinton Administration to 'provide incentives to act quickly', had featured in the *Wall Street Journal* on Day One.

I spent a long time over my sushi wondering what the climate was like in the White House. I thought back to the time I had spent with the earnest and patently sincere Senator Gore in 1991 and 1992, discussing climate change. What must he be thinking now? I could only conclude that the man I had met then would be feeling torn and guilty. But what would five years as vice-president of the United States do for core values? On that, I could only speculate.

I read a poignant critique from a provincial journal, the *Milwaukee Journal Sentinel*, written by a leading figure in a student organization called Campus Green Vote. She had once been a canvasser for Gore on the campuses. Now it sounded as if she hated the man. Her article held a nasty sting in its tail. 'Think of the young people who will inherit the problems you postpone,' she finished, 'and, by all means, reread your book.'

Even in Gore's conservative home state, the heat seemed to be on him. An editorial in *The Tennessean* concluded as follows: 'What the United States should not cede is its leadership on this important issue which has the most serious consequences for the planet. Gore used to be kidded for his interest and concern about such an arcane topic. Now he has the chance to be taken very seriously indeed.'

On Day Three, the war of words between environmental groups and business lobbyists began to heat up. Friends of the Earth had hit on the idea of a 'Scorched Earth Award'. A trophy consisting of a bowl of smoking dried earth would be awarded to the worst carbon criminals at a ceremony on Friday. In the interim, delegates would have the opportunity to vote for their preferred candidate. The short list included Exxon, Mobil, Shell, General Motors, Ford, Tokyo Electric Power and the Global Climate Coalition. Tony Juniper, FoE's ebullient campaigns director, told me that votes were already flooding in.

In the large room set aside as the business NGO centre, I circulated freely with the representatives of the shortlisted companies and the rest of the business lobbyists. I had yet to attend my first morning coordinating session, and now Clem Malin let it be known, via an International Chamber of Commerce official, that he wanted to see me before the session. I had already been told by a friend in the secretariat that the ICC had complained about my presence as a business NGO. The secretariat had rebuffed them. I was now a genuine business representative, the secretariat said, and the ICC would simply have to put up with it.

I found Malin working on a document in the meeting room for business attendees. It transpired that he wanted to tell me he was concerned that if I turned up at the business coordination meetings, others might simply stop coming. This, he said smoothly, would be a shame.

What did they think I was going to do, I asked, froth at the mouth and disrupt proceedings? These days I was the managing director of a limited company. I chaired an industry taskforce in the UK on which blue-chip energy companies, insurers and banks served. I had every right to be treated as a regular business representative.

Malin looked as though he was thinking of saying something more, but didn't. Our brief conversation, I gathered, was over.

Out and about in the conference centre, the extreme fringe among the hundreds of industry lobbyists was already hard at work. J. R. Spradley, representing the Edison Electric Institute, was prominent among them. In all the seven years I had known J. R., the arguments I found so persuasive had evidently not registered to the slightest degree in his consciousness. He still maintained there was no proof of a global warming threat, and still defended his case with a stubborn pride. With him, I felt as if I were discussing religion, not science. With his 'nice suit, sharp haircut, and tassled loafers,' so the *Washington Post* wrote, J. R. was 'the kind of guy who makes environmentalists crazy'. The *Post* then offered the following immortal quote from him to illustrate the point. 'How many people were following Moses when he started? And there was only one guy saying the earth was round in the beginning. It's nothing to be ashamed of.'

That evening, at a reception, I came across the court jester of the professional sceptics, Fred Singer. Something impelled me to check out the current status of his mental global warming map. 'Aren't you worried,' I asked him, 'about how you will be portrayed in the history books, say twenty to thirty years from now?'

'I never think about it,' Singer replied without pause. He was interested in data, he said. He suggested that when he next visited Oxford, he and I have a private discussion about data. A permanent lazy smile sat on his face.

I embroidered my theme, using the medical opinions that had been aired by World Health Organization experts the previous day. At a press briefing in the afternoon, four leading medics had concluded that at least 8 million lives could be saved if Kyoto succeeded in adopting targets. Earlier in the week, over 400 physicians from 30 countries had signed a 'Medical Warning' in the *New York Times* supporting action. They included Nobel laureates and editors of medical journals including *The Lancet*. Didn't it worry Singer that the consequences, if he was wrong, included millions of avoidable deaths?

A momentary impatience disturbed Singer's maddening smile. 'Those guys aren't scientists. They are propagandists.' Without elaborating, he switched to a cloying attack. 'You know, Jeremy, I'm glad to see you are consistent. I had thought you had just opted to take the Greenpeace money. But it seems that you are intellectually honest. Misguided, but intellectually honest.'

Struggling to keep my own fixed smile in place, I decided I had heard enough. 'As for me, Fred, I believe you should be assigned your appropriate place in history. I can promise you I will do my best to make sure you get it.'

On Day Four I made my inaugural appearance at the daily business NGO coordination forum. In the meeting room adjacent to the business NGO centre, some fifty men and women sat around an oblong table waiting for Clem Malin to open proceedings. I took a seat, my pulse rate irritatingly high. As far as I could see, nobody was heading for the door.

Next to me, an American I did not know offered an aside to one of his colleagues. 'I didn't know we had socialists in our meetings.'

Malin called the meeting to order. Before beginning, he said, he wanted to introduce a new member to the group.

'Welcome Jeremy to the league of carbon criminals.' And then, to the group, 'I guess he's discovered there is more money on our side.'

Grins around the table.

Michael Jefferson, ex-Shell and now representing the World Energy Council, addressed himself to Malin. 'Are you suggesting that *you* are unmotivated?'

'Oh no,' Malin laughed.

It was my first clear view of what I knew already. This was not a monolithic group.

Malin steered the gathering through a routine exchange of information about the directionless events of yesterday, and the conference agenda for the day. I heard nothing untoward during this, save a reference to China 'wanting it for free' in a discussion about technology transfer discussions. Indeed, it was remarkably like the environmental NGO coordination meetings I had attended for Greenpeace in the past, though if this particular session was anything to go by, with a generally far less sophisticated level of political analysis from the participants who spoke. I did not attach too much significance to this. I knew that the serious black-hat strategizing would go on within the privacy of each of the carbon club's individual business groups. After all, in this wider coordination forum, not counting my own presence, there were such questionables as Michael Jefferson, plus the representative of the European arm of the Business Council for a Sustainable Energy Future, Paul Metz, and even – these days – BP's Klaus Kohlhaus.

Jefferson, a dapper and knowledgeable Englishman, came up to me afterwards and apologized on Malin's behalf. 'Completely out of order,' he said.

Meanwhile in the negotiations, frustrations were now bubbling over into open acrimony. Ambassador Gremagna of Luxembourg, speaking for the EU presidency, accused the host nation of abusing its position by relaunching and leading an offensive on the EU bubble. He also attacked the idea that the EU should take on tougher targets than Japan and the USA. That would be politically impossible to sell to the European people, he insisted. Most pointedly of all, he alleged that Japan was 'deliberately not understanding' explanations.

Japan's senior negotiator, Tashiaki Tanabe, dismissed the European outburst as an effort to win public sympathy. Behind closed doors, he said, serious horse-trading was going on.

But the Global Climate Coalition seemed sanguine. John Grasser of the US National Mining Association, a Coalition representative, told the press that the Kyoto Climate Summit was heading nowhere. 'We think we have raised enough questions among the American public to prevent any numbers, targets or timetables to achieve reductions in gas emissions being achieved here,' he said. He was nothing if not frank. 'What we are doing, and we think successfully, is buying time for our industries by holding up these talks.'

On Day Five the insurance companies appeared at the Conference of Parties for their customary one day a year. At my invitation, Tessa Tennant of NPI and Andrew Dlugolecki of General Accident attended the morning business NGO meeting. So it was that the first insurers saw pretty much the full brigade of leading carbon-club lobbyists arrayed.

Tessa and Andrew were immediately offered a revealing glimpse of the bottom line. Clem Malin began by telling the gathering about a rumour that had been circulating the previous day, to the effect that the USA was prepared to drop its target from a freeze in emissions to actual cuts. You could see the indignation in the body language around the table. This had been roundly denied at a press briefing by the delegation the previous evening, Malin said, his relief clear.

When the meeting closed, I took the insurers up to meet the Texaco man, in Andrew Dlugolecki's case for the second time. Incredibly, after Berlin, Malin had felt the need to fly across the Atlantic to Edinburgh to visit General Accident. Malin's mission, so Andrew had told me, was to persuade him that General Accident's strong advocacy of action on greenhouse emissions was unwarranted. Unsurprisingly, Malin had drawn a comprehensive blank. But now, as the two talked, I had the clear impression that Malin was no longer worried about the presence of the insurance companies. His manner was relaxed, and I sensed he knew there was little weight behind the insurers' presence; that they had little support from their boards, and would disappear the next day, not to be seen for another year, with little or no follow-up to trouble him.

Out in the main stadium-sized hall, Friends of the Earth had counted the votes for the Scorched Earth Award, and were preparing to present the trophy. The Global Climate Coalition had won.

Amazingly, it seemed, the executive director of the GCC, Gail MacDonald, had agreed to come and accept the award. Perhaps she figured she could turn the media attention to her advantage in some way.

A big crowd gathered. Tony Juniper, as master of ceremonies, explained to newcomers what the award meant, and how the voting had gone. A battery of press photographers hovered. The trophy smoked in front of Tony as he spoke.

One of his colleagues whispered in his ear.

'Ladies and Gentlemen, we have a disappointing announcement,' Tony said. 'The Global Climate Coalition have seen fit not to come and explain why they think it is morally defensible to wreck the planet.'

The environmentalist had a reserve card, of course. 'Before we go to them to present the award, we have other business spokespeople to offer you a few impressions.'

Tessa Tennant then stepped forward, and in a small but significant way made a bit of history. She represented a large British pensions and life assurance company, she said. And her company, NPI, had an important and ground-breaking announcement to make this day in Kyoto. 'We are calling for governments worldwide to introduce global health warnings on all oil and petrol advertising.' This was a precautionary measure which all governments could readily take, she continued. It was widely recognized that transport emissions were a significant and increasing source of various pollutants. 'Let's focus on this and get the message across to people in their daily lives. This is about managing environmental risk. An informed public is essential to the success of action plans to reduce man-made emissions to "liveable" limits.'

Next, briefly, it was my turn. I offered a few rhetorical sound bites from the Industry Solar Taskforce Statement, casting them as a business counterview to the statements and actions that had won the GCC this award.

Tony Juniper slowly headed in the direction of the business NGO centre, holding the Scorched Earth Award in front of him like a sacred offering. A scrum of photographers and cameramen retreated backwards before him as he advanced with the smoking bowl.

I followed in the crowd at a distance.

At the closed door of the centre, the Coalition had positioned one of their young gophers as a guard. This all-American-looking figure proceeded to tell Tony that he could not enter the door of the business NGO centre. Cameras whirred and clicked as he spoke.

This was a shame, Tony retorted. He would just have to leave the award outside.

Yes, said the gopher dismissively, motioning at the other side of the corridor. 'Put it over there.'

I felt an impromptu devil rise in me, and my mouth was open before I knew it.

'Tony, I am a business NGO. I'll deliver it for you.'

I stepped forward from the crowd, and Tony passed me the bowl, grinning from ear to ear.

'You can't stop me going in there,' I told the gopher quietly.

Another suited figure stepped forward from the crowd. It was Paul

Metz, the debonair Dutch executive director of the European Business Council for a Sustainable Energy Future.

'Jeremy, I will assist you in delivering the award.'

And we opened the door to the business room, holding the bowl between us. The sound of camera shutters merged into a continuous rippling whir.

I looked at the incensed faces of the dozen or so Global Climate Coalition people within.

'Please,' I said, 'allow us to present you with the Scorched Earth Award.'

The insurers held their press briefing shortly afterwards, and the media questions soon homed in on the issue of investment. The *Guardian*'s man in Kyoto, Paul Brown, was particularly direct. 'Isn't it,' he asked, 'really rather simple?' Don't you just threaten to disinvest unless an oil company starts wholesale investment in solar and other renewables? That could solve the global warming problem in short order. 'So, have you made any decision not to invest in Exxon yet?'

The insurers were predictably cagey about addressing specific companies, but I heard some of the strongest public comments on this subject yet.

'What we are saying to individual companies in which we have large shareholdings,' Andrew Dlugolecki responded, 'is that we are analysing their strategies on the environment. We will be looking particularly at companies in the energy, transport and tourist industries. If they have not got a good strategy for dealing with climate change, they will not be a good investment.'

Yes, agreed Ivo Knoepfel of Swiss Re. 'Oil companies need to reclassify themselves as energy companies and get into the future.'

The press briefing ended, and immediately afterwards the insurers held a two-hour seminar to explore details. At this meeting I could not help airing my disappointment with the industry. I had come to a rather depressing conclusion, I announced in the discussion. It seemed to me now that before climate change became a price-sensitive issue there would have to be a big catastrophe to wake the market up.

Andrew Dlugolecki responded. The essence was that he wasn't sure he agreed. We didn't need to be quite so pessimistic.

At the end of the seminar, Munich Re's technical chief, Gerhard Berz, came up to me. He too wasn't sure he agreed either, he said, smiling ruefully. 'I think it will take two to three major disasters.'

With Day Six came the last chance for negotiators to make progress before ministers arrived. Sunday was nominally a day of rest, and the high-level segment of the summit was to begin first thing on Monday.

During the day, Chairman Estrada repeatedly expressed frustration at the lack of flexibility being shown. 'We are late in the process. We are late in the discussions,' he said. 'I really would like to invite countries, especially Annex 1 countries, to show flexibility.'

The discussions went on long into the night. But as the hours went by, delegates who from time to time left the hall to update waiting NGOs reported no sign of a compromise emerging.

Meanwhile, the confrontation between developing and developed countries deepened. New Zealand proposed that the developing nations make a binding commitment to limit growth of their emissions after 2015. The USA let it be known that the New Zealand proposal enjoyed 'conditional' US support.

Ten minutes before New Zealand made this intervention, which of course was bound to inflame many of the developing countries, Don Pearlman was overheard by an environmental NGO asking a delegate whether 'the bomb had dropped yet'.

Could his influence even extend to stooges in countries like New Zealand? Who could know. Pearlman was constantly to be seen hovering outside the closed sessions of the negotiations, cigarette in hand. He conducted earnest conversations with many diplomats, some of whom clearly came out to brief him – or receive instructions – and who then went straight back in again. Without observing the secretive man round the clock, it was impossible to work out whom he might be working with, apart from the obvious players in the Saudi and Kuwaiti delegations. Yet it seemed inconceivable that he would put all his eggs in that one basket.

From time to time, TV crews or press photographers would train their cameras on him, hoping for an incriminating shot. Environmental NGOs were constantly pointing him out to newcomers to the negotiations. But they did not have the patience to linger for long, and anyway these days Pearlman spent a lot of time looking round to check who was watching him as he went about his work.

Sunday dawned with blue skies, and there was no question of my being one of the many for whom this would be a day off in name only. I asked where I could find the most beautiful spot in town, and was directed to the Kiyomizu Temple, in the woods above the city. I set off through the narrow streets.

I found a temple complex dating from the eighth century nestling in still-autumnal mixed woodland. It was a place of serenity, where despite the crowds of visitors I knew I could linger for hours and recharge. As I approached the magnificent wooden structure, its great overhanging roof curling upwards at the corners in the classical ancient Japanese style, I was greeted with the incongruous sight of a sizeable solar PV array in the yard in front of the temple.

Sure enough, it was a Greenpeace demonstration, and – I was astonished to see – a permanent one. I learned later that the monks had welcomed the chance to make a statement about global warming during the Kyoto summit. My former colleagues had had no problems persuading them to juxtapose the future with the past in this way.

But as I looked, I heard a familiar rasping voice behind me. 'What are you doing ruining this beautiful place with your solar panels, Jeremy?' It was Mobil's master of propaganda, Leonard Bernstein.

'And what are you doing ruining this beautiful day for me by turning up in it, Leonard?'

Japanese buddhists were not the only religious body making a point about global warming that day. The World Council of Churches, representing a very broad sector of religious organizations, chose Sunday to issue their statement in Kyoto. They called on governments, in no uncertain terms, to support the AOSIS protocol. As the first week ended, it was pretty clear which side the deities were on.

I was wise to take the day off, it turned out. By the end of the first week of the summit, around thirty delegates had been admitted to hospital with dehydration and exhaustion. On Sunday, Tony Juniper joined them. The irrepressible Friends of the Earth director collapsed and was put on a saline drip overnight.

*

On Day Eight the conference centre was packed. With the arrival of ministers had come a further influx of journalists and NGOs: 5,500 journalists were now registered. This was to be the day that Al Gore would say his piece, and the serious endgame would begin.

The mêlée outside the main hall was a sight to behold. Getting a seat in there for Vice-President Gore's speech was clearly going to be impossible, so I found a chair by one of the big TV screens. Coffee in hand, I settled down to wait. Large crowds soon built up around each screen.

The first major speech of the high-level segment was by Japanese Prime Minister Ryutaro Hashimoto. Climate change, he said, was a 'direct threat to the future of humankind'. But he went on to give the latest in the lengthening catalogue of national leaders' speeches, over the years, in which rhetoric about the threat far outstripped sincerity and urgency about the response.

The second of the three heads of state in Kyoto was Costa Rican President José Maria Figueres, who spoke on behalf of many developing nations. 'My friends from the North,' he concluded, 'the ball is in your court.'

The president of Nauru, Kinza Clodumar, elaborated from the perspective of the small island states, and he put their case – as so many from the islands had over the years – with emotion and aplomb. 'The wilful destruction of entire countries and cultures with foreknowledge would represent an unspeakable crime against humanity. No nation has the right to place its own, misconstrued national interest before the physical and cultural survival of whole countries. The crime is cultural genocide. It must not be tolerated by the family of nations.'

Clodumar continued in this vein for some minutes before coming to the bottom line. 'President Clinton promised that the US would bring to Kyoto a pledge for significant future reductions.' He turned to his left to address the seated Al Gore directly. 'Vice-President Gore, we await your announcement with bated breath.' He ended with a cross between a plea and a prayer. 'Let us create a Kyoto Protocol that we can show proudly to our children. Let us take action, effective action, prompt action, here in Kyoto, without reservation, without delay, for now and for ever.'

As a thunderous ovation greeted the Nauru leader's speech, a smiling Al Gore came into camera view to pump his hand, and clap a hand on his arm.

The vice-president began his own words with a routine global-warming comment. 'The trend is obvious,' he concluded. The challenge was to find out whether and how we could change. None of the proposals on the table would solve the problem. They were all a first step. That first step must involve realistic and binding targets.

The USA had listened to the developing countries, Gore said. 'We do not want to founder on a false divide.' The USA had listened to its developed country partners too. 'You have shown leadership and we are grateful.'

But the US proposal, Gore emphasized, was serious. 'It involves a 30 per cent reduction from what would have been.' He reached the bottom line: 'After talking to President Clinton a few hours ago, I am instructing our negotiators to show increased flexibility.'

As soon as the speech finished, NGOs and journalists broke from their ranks around the TV screens to offer and record reaction. A line of four chosen environmental NGO representatives from Africa, Asia, America and Europe formed up in front of a battery of TV cameras. The condemning sound bites began. 'I don't know who we heard, Al Gore or Al Bush,' raged Atiq Rahman from Bangladesh.

Extract copies of the *Congressional Record* from 7 May 1992 had been passed out en masse earlier. They bore out Rahman's barb with brutal precision. Speaking on the eve of the Rio Earth Summit, Al Gore had had this to say of President George Bush and his moral responsibilities for the climate convention: 'It is about far more than hopping on a plane for a quick photo opportunity . . . It is not about trying to pull the wool over people's eyes and pretending to be doing something when actually nothing is being done. It is about leadership, it is about courage, and the president is exhibiting neither of these qualities. It is about embracing a perspective that extends well beyond the next election.'

To emphasize the hypocrisy that so many NGOs felt Gore guilty of, Tony Juniper – now off his saline drip and back in action – read passages from *Earth in the Balance* as a lament while TV crews filmed him.

Reaction from official Europe was mixed. The Danish environment minister, Svend Auken, told journalists he thought the speech was 'diplomatic climate fraud'. John Gummer was just as forthright. 'Rubbish,' he snapped at TV cameras, speaking about Gore's claim that a 30 per cent cut from projections was as good as any nation proposed. 'By his formula, Britain is offering a 50 per cent cut.' But Peter Jorgensen, the European Commission spokesman, held a different view: 'Vice-President Gore has given us a window of opportunity, and we don't get many of those.'

This was the view of the big American environment groups, who invariably tended to be less critical than their international counterparts. The Environmental Defense Fund thought the speech a significant step in the right direction.

For myself, I viewed both the speech and the strategy as, at best, cruel brinkmanship. That view would be contingent on a protocol being agreed.

The Global Climate Coalition press release appeared almost before Gore had left the podium. It beggared belief. 'The current White House position will result in a 30 per cent reduction in energy use nationwide. There is no available technology to accomplish that goal.' The second sentence was underlined.

Here were two grotesque examples of Global Climate Coalition distortion in two concise sentences.

Senator Chuck Hagel of Nebraska, one of the six senators at the summit, put out a statement on behalf of the carbon club's legislative flankers. 'The Senate has spoken,' he stormed. 'We will not support any treaty that does not require the developing countries to sign on to the binding commitments that we're asking of the United States.'

The Senator's scaremongering echoed the Global Climate Coalition's exactly. 'Entire industries will leave the United States for countries that won't be bound by this treaty. For the first time in American history we would be giving an international body the authority to limit and regulate our economic growth.'

That same morning, an advertisement appeared in the *Washington Times* plunging the tenor of the conservative case into new depths of distaste. It showed a photo of the Japanese surrendering at the end of the Second World War. 'America has signed many treaties . . . but never a treaty of surrender. But that is what could happen in Kyoto.'

*

Vice-President Gore gave a press briefing soon after his speech. 'President Clinton and I will be hard at work behind the scenes,' he promised, 'telephoning presidents and prime ministers and asking them to deal with positions that their negotiating teams are taking when we believe that they're not helpful.'

He was asked the obvious vital question. What did he mean by a new flexibility from the US negotiators? 'I'm going to leave the specifics to them. I think that's the right way to do it. But the president and I have been specific with them about their instructions, and, in due time, you will see exactly what it means.'

And indeed, by the end of the day European diplomats were reporting a greater willingness by US diplomats to plug loopholes in the details of trading and joint implementation. More than that, they reported that in his fourteen hours of consultations, Al Gore had not ruled out the possibility that the USA would raise its target for curbing emissions.

On Day Nine both the World Business Council for Sustainable Development and the Business Council for a Sustainable Energy Future would be giving press briefings. I went to the World Business Council event. Among the row of executives on the platform was BP's Klaus Kohlhaus. BP had put the Global Climate Coalition far behind it. Today it represented the Union of Industrial and Employers' Confederations of Europe. 'UNICE and industry in Europe are prepared to support targets,' Kohlhaus announced. 'Industry is actually contributing and delivering.' I had never before heard Klaus Kohlhaus speak in this vein. What was more, he spoke for 130 companies in 35 countries in 25 industrial sectors. He went on to offer the BP view. 'My company has come up with a strategy on climate change because we believe there is a business opportunity for solar in the years to come.'

The contrast here with the US oil companies was total. I wondered what Clem Malin would have thought about the performance had he been there to see it. After all, Texaco was a member of the World Business Council. Others speaking at the event, including WBCSD President Bjorn Stigson, echoed the BP man, and though they admitted that the spread of the opinion within the WBCSD did not allow a consensus statement about targets, it was clear that Texaco had been marginalized.

During questions, I took a turn at the floor microphone. I recalled the Business Council's press conference in Rio, seven years ago, and the performance given then by the Italian oil company ENI on behalf of oil and gas interests in the Council. It seemed to me that the companies here today had come much further than governments in the interim, I volunteered. Would the panellists agree? And if so, why is it that the companies in the Global Climate Coalition hadn't enjoyed the same evolution?

Stigson responded first. He agreed. They were doing this because it made business sense.

Klaus Kohlhaus guardedly addressed the last part of the question. There were different views in industry, he said. But BP had been listening to its customers and the public perhaps more than its counterparts had in the USA.

The US and European arms of the BCSEF made a joint intervention in the high-level session, where speechmaking was still underway. This group, representing Enron and other renewables and efficient-energy interests, gave a firm alternative view to the GCC's incredible statement about technology not existing for reductions in the projected emissions increases. 'We all know with confidence,' US Executive Director Mike Marvin said, 'that appropriate steps to respond to climate change – based upon the efficient and clean use of energy – will lead to long-term, worldwide economic growth. An early and meaningful reduction target for Annex 1 countries will help convince developing countries that a less carbon-intensive economy is viable.'

In the late afternoon, the European Union held a press briefing. The Environment Commissioner, Beau Bjerregard of Denmark, was upbeat. 'I still hope for an ambitious target,' she said. 'I haven't given up hope. The Americans are serious and trying hard. We are still negotiating with them, and it is still too early to say what the outcome will be yet.'

German TV asked the question that was on everyone's lips. 'Is there any sign of the increased flexibility that Vice-President Gore talked about yesterday?'

Bjerregard smiled. 'Yes,' she said.

By evening the waiting around and constant rumour analysis had led many to the sake bar in the foyer of the main hall, myself and J. R. Spradley included. I stood there talking to him, watching Don Pearlman tirelessly working the delegates in the distance.

I had come by a copy of a letter intimating that the Edison Electric Institute had been guilty of suppressing a consultancy study showing that the economic costs of US emissions reductions were nowhere near as high as industry hoped to show. Californian Representative Henry Waxman had written to the Institute – J. R. Spradley's employer and one of my funders – demanding to know if this, as reported in *Air Daily*, was true. I now asked J. R. if indeed it was.

'You should never believe everything you read. You know, people take half a piece of information and stick it in the press.'

'Oh, the Global Climate Coalition never takes half a piece of information and sticks it in the press?'

'I dunno,' Spradley replied, slugging back a cup of sake in one. Then he grinned at me. 'Maybe once.'

At the long-awaited US press briefing, the potential deal finally took some shape. 'Our proposals include real reductions below 1990 levels in the 2008–2012 timeframe,' Undersecretary Eizenstat told the press. 'This represents significant movement on our part.' But, he emphasized, the deal was contingent. It would have to include all six greenhouse gases, appropriate carbon sinks, flexible market mechanisms such as trading and joint implementation for credit, and meaningful participation of key developing countries.

Brazil had come up with a proposal which met with American approval, it seemed. It involved a so-called clean development mechanism that would encourage investment in new energy savings technology in the developing world.

But Eizenstat's concession came with a warning: 'We do not sense the urgency on the part of many countries that is necessary given the lateness of the hour.'

The Global Climate Coalition rushed out its predictably rabid response: 'President Clinton said the United States should walk away from a bad deal. This is a terrible deal. The United States should not walk away. It should run.'

But this was an attempted deal evidently now being driven by the USA.

The environmental NGOs' evening press briefing was so packed that hundreds viewed it from the gallery above the Event Hall, looking down into the roofless NGO centre.

'This night might be the most important night in the history of the climate,' said the Worldwide Fund for Nature's spokesman.

'The next twenty-four hours are the most important in the history of this planet,' said the Climate Action Network's spokeswoman, not to be outdone.

On the last day, 10 December 1997, a mood of grim indignation hung in the air at the morning industry NGO coordination meeting. A deal was still a possibility. This was not the plan at all.

Clem Malin provided the carbon club with their main ammunition for the day, a letter from the Senate majority leader, Senator Trent Lott, to Senator Hagel in Kyoto. Malin read out the key part: 'I have made clear to the president personally that the Senate will not ratify a flawed climate change treaty.'

Photocopies of cartoons from the conservative press were also handed out. One, from the *Orange County Register*, showed a towering militiaman labelled 'U.N.' and 'Global Warming Police' wagging a finger at the Statue of Liberty. 'Do you have a permit for that torch?' asks this fascist-like figure.

All morning the small contact groups of negotiators with final responsibility met. Like the day before, there was little for the majority of the 10,000 delegates to do except wait, and gossip. Only those with proxy ears and voices inside the secret meetings, such as Don Pearlman on one side and Greenpeace policy director Bill Hare on the other, were able to attempt to influence things at this late stage.

The final Conference of the Whole, where the deal would be done or ditched, was due to start at 1 p.m. But one o'clock came and went. The new target for the start was set at 6 p.m. It seemed that a group of eleven countries was cutting the final prospective deal: the USA and Japan for JUSCANZ; the UK, the Netherlands and Luxembourg for the EU; and for the developing countries China, India, Brazil, Colombia, Samoa and Tanzania.

In the vacuum, environmental NGO representatives with dark circles under their eyes staged impromptu demonstrations in the Event Hall. Anti-American sentiment was running high, and one of these demonstrations – approaching if not quite matching the xenophobic paranoia of the *Orange County Register* cartoon – featured an American flag, a figure in a hood, some oil-industry corporate logos and a hangman's noose.

At half past six the Conference of the Whole came to order. Deep crowds gathered round the TV screens.

Ambassador Raul Estrada opened proceedings. The final draft text, he said, would be presented to delegates when the papers were ready, and the final session would begin at eleven o'clock.

We now knew for sure, as if we hadn't before, that this was going to be an all-nighter.

Estrada spelt out his plan. The Conference would begin by considering Article 3, on commitments, and then go from 1 to 28 in sequence. Thereafter there would be two annexes: 'A', listing the gases – all six – to which the protocol would apply; and 'B', where the differentiated targets to be undertaken by the developed countries would be listed. This text would be considered on pretty much a take-it-or-leave-it basis, he made clear. It would be submitted only for improvements. If it could be agreed, it would then go to a brief formal Conference of the Parties chaired by Japan, where it would be adopted.

'If we can reach a binding agreement,' the chairman said dramatically, 'this day will be remembered as The Day of the Atmosphere.'

If not, he left unsaid, we would face further years of unfettered fossil-fuel profligacy. Worse yet, we would know the surreal despair of wondering whether governments would ever prove themselves capable of tackling – even beginning to tackle – a problem which threatens the future of our species.

I left to have dinner with a group of my old colleagues from environmental groups. As we ate, mobile phones rang incessantly with incoming messages from last-minute contact groups considering the draft text. A delegate reporting from inside the G77 and China group said that by the end of that night there might no longer *be* a G77. Most countries were willing to accept any compromise deal at this stage, but China and India were filibustering on trading and joint implementation.

As we gulped coffee a call came from a member of an American NGO. He had just met Eizensat, who had told him the talks were on the verge of collapse because China was vehemently opposing emissions trading.

We need not have bothered rushing, because when we got back to the conference centre it was clear that the final Conference of the Whole would not be starting at eleven.

I wandered round the conference centre as the thousands waited. The hall where the Conference of the Whole was to meet was a bizarre spectacle. The final session would be completely open, and the back wall was already lined with a bristling battalion of TV cameras. CNN would be running live coverage.

Famous broadcasters drifted through the crowds, mobile phones in hand. In the galleries, NGOs had already taken all the available seats, plus much of the standing room, in order to be able to see the drama live. I knew that far more comfortable options existed around the big TV screens outside, but I had resolved to watch the denouement in the vicinity of Don Pearlman, so that I could see his response as the events he had done so much to shape finally unfolded.

To my good fortune, Pearlman opted for a guaranteed seat by a TV screen. I took a seat close by, behind a now-deserted UN information desk. I put a 'closed' sign on it, and settled down for the night. My eyes were already red.

Most of Pearlman's hardline colleagues were with him. Leonard Bernstein of Mobil sat on his left, with Gail MacDonald of the Global

Climate Coalition two seats along. Nearby, in a clutch of other foot soldiers, sat Clem Malin of Texaco, Brian Flannery of Exxon, Constance Holmes of the US National Coal Association and John Schiller of Ford. Here, within the thickness of a coal seam or the diameter of an oil wellhead, were the main orchestrators of the fossil-fuel industries' prolonged rearguard action.

The thousands milled as the telephone calls from world leaders jumped around the planet, and the final draft text was prepared. Just after one o'clock, flustered UN officials appeared in the conference hall with piles of hot photocopies of the draft Kyoto Protocol. They began handing them out, and I was lucky enough to get one of the first. It was just as well, as a writhing scrum of people, arms extended and hands grasping, gathered within seconds.

I turned straight to Annex B. The numbers were not yet there, but the rumour was that the USA had agreed to a 7 per cent cut, with Japan opting for 6 and the EU 8.

Of course, the conditions had first to be agreed.

I scanned quickly through the rest of the document. Trading was covered in three paragraphs of Article 3 on targets and timetables. Joint implementation was in Article 6, and the whole of Article 9 was a carefully worded manifesto for how developing countries could take on binding targets if they wished to do so, at any time. They could take on a level of emissions limitation, or reduction, meaning that if they wanted to they could simply commit at this stage to limit the rate at which their emissions increased.

And now the final session of talks began. With 28 articles and two annexes to shepherd through, and some seven hours before the UN's translators would have to leave Kyoto, Chairman Estrada kicked off with Article 3, as he had said.

He made rapid progress until he came to the paragraphs on emissions trading. China then objected. This was new language, their ambassador said. The old article on this subject had been dropped.

India immediately joined them.

Saudi Arabia, quick as ever to exploit any sign of an embryonic divide, egged them on.

Brazil supported the trading language. Russia spoke indignantly about what it saw as its 'sovereign right' to trade. The discussion ebbed and flowed.

The minutes became an hour.

Estrada appealed for delegates to simply accept the text. They had discussed it at length in AGBMs, he stressed, and everyone knew that the commitments proposed by some parties were contingent on this language.

The Americans now spoke up. 'We know full well we are the world's largest emitter,' Eizenstat said, reprising the old refrain which had been thrown at them. The USA knew all about the responsibility to show leadership, he emphasized. 'This discussion is the most important in the history of the global climate issue.' And he wanted to be clear. The USA had moved from its opening position of a freeze by 2010 to 'very deep cuts'. Others had moved too – the EU, New Zealand, Canada. But the USA required innovative options to meet these targets. 'I urge in the strongest terms, now that these historic commitments are about to be made, not to deprive us of the means to meet them. The eyes of the world are upon us. We have the opportunity to do something great.' But the trading mechanism, he said, was needed to do that.

Canada and Japan spoke up in immediate support.

So too did Samoa, for AOSIS. The island nations were now desperate for any deal.

China came in again. 'What are the rules of the trading game? This question has to be addressed before it can be put into a legally binding instrument.'

The language in Article 3 deferred rule-making and guidelines for verification for later agreement by the Conference of Parties.

India supported China. The compromise wording was no compromise at all, and now the Indian ambassador proposed entirely fresh wording.

Uganda supported both China and India, and Saudi Arabia stepped in to support all three.

I saw with shock that three hours had now elapsed. I felt a bitter impotent anger rising inside as I watched.

With the clock approaching 4 a.m., Chairman Estrada appealed for sanity. 'We are about to blow any chance of an agreement,' he warned. 'It has long been understood that trading should be included.

One side has shown some flexibility, the other not. We need more flexibility.'

Soon after that, he called for a five-minute recess. His intention, to knock heads together in private, was clear.

I watched Malin of Texaco, Bernstein of Mobil, and Flannery of Exxon laughing together. The negotiators were nowhere near the issue of maximum contention yet. I looked around, and marvelled that at that point – with perhaps 3,000 journalists in the conference centre – not one was photographing or filming the moment of carbon-club triumphalism.

Pearlman, strangely inscrutable, did not join in. I marvelled that no OPEC diplomats had been out to him for instructions, as they normally did.

The recess turned into 30 minutes, and I felt then the despair of knowing there would be no protocol.

Chairman Estrada came back with a compromise. Trading would be included, covered in a fresh article, '16 bis' in arcane UN treaty language, but the details would be worked out later. He raised his gavel, and hit the desk to signify adoption of the article.

Most of the diplomats applauded.

One down and 27 to go.

'We only have two hours in all six languages,' Estrada warned. He then adopted two articles, four on compliance and five on methodologies, within as many minutes. The clock read 5.20.

Immediately, Kuwait tried to stall the next article. Estrada brushed them aside. 'I don't see any possibility of consensus on that. So Article 2 is adopted.' He hit the desk with a loud bang of his gavel.

Kuwait came back. 'This protocol affects our future livelihoods. We cannot just push things through.'

'Thank you. Let us go to Article 6. I see no objections.' Bang. 'It is adopted.'

Clapping and now laughter filled the hall.

Two more articles received the same treatment, and hope rose fluttering inside me.

Stony faces now stared at the screen from the carbon camp.

The conference hall had by this time taken on the appearance of a refugee camp. Less committed diplomats and journalists slept in rows against walls, or slumped in chairs, oblivious to the drama.

But now we came to Article 9, setting out terms for participation by developing countries. Saudi Arabia asked for its deletion, immediately supported by China and India. There would be no gavelling this through, and debate welled again, with the USA, Russia, the island nations and Argentina insisting the Article stayed.

Six o'clock came, and Estrada appealed for countries to make only brief interventions. 'We are really so close to the end of our facilities.'

Soon after, the chairman took a desperate decision. He could not see any chance of consensus, he said.

'We will delete this article.'

This would be the moment for a US walkout, and a carbon-club victory. India and China would be blamed, at least as much as the Americans, if not more.

But the USA sat tight.

I watched the carbon club for reaction. There was none. They were all awake, which was more than could be said for many environmentalists at this point. But the grim faces stared at the screen torpidly. Nobody conferred.

They must have been contemplating the grim contours of their fight in 1998. The one article on developing-country participation, mildly worded as it was, had now gone from the protocol. This omission would have to be the centrepiece of their campaign to prevent ratification.

Estrada launched himself into a further phase of rapid gavelling.

We came next to joint implementation, Article 13. Another pedantic debate erupted over how the already differentiated targets could be tinkered with by offsetting emissions achieved overseas. The protocol on offer in Kyoto was never going to be perfect, and now the more selfish among the industrialized nations defended their partial get-out clauses.

At 6.50, copies of the *Daily Yomiuri* arrived in piles. 'DEAL STRUCK IN KYOTO TALKS' the headline screamed. Channel Four's correspondent walked past talking worriedly into a mobile. 'This thing is falling apart,' he told his newsdesk.

By dawn, they were still less than halfway through the articles.

Finally, Chairman Estrada gavelled through a compromise on joint implementation.

Nobody objected. Nobody walked out.

The major points of contention had come and gone. But it was so, so late.

I watched with rising hope as Estrada once again began rattling off articles.

He came to the article on entry into force. The text stipulated that the protocol would come into force after it had been ratified by parties responsible for at least 55 per cent of the carbon dioxide emissions in Annex 1 countries. Canada proposed 65 per cent. No, said Estrada, that would give a veto in principle to one country (the USA). Japan and Australia proposed 60 per cent. The Marshall Islands supported the original.

'There is no consensus,' Estrada said impatiently. 'We adopt the original.'

Two more zipped by. It was 7.35, and he had two final articles, plus the two annexes, to go.

One of the articles involved the touchy issue of sinks, and another round of carping interventions began. Argentina tried to reintroduce the article on developing-country participation. No, China said stridently. That cannot be done once the chairman's gavel has fallen.

Estrada allowed the discussion to ebb and flow for a while, and then made his decisions, and gavelled through the compromise language.

Nine a.m. had arrived. The exhausted interpreters could not stay a moment longer.

The equally exhausted negotiators endured a lengthy wait as the draft of Annex B was prepared. At this point, Don Pearlman made the only mobile phone call I had seen him make all night.

Finally, the single sheet of numbers arrived. The clock was approaching 10 a.m.

The USA had indeed agreed to a 7 per cent reduction target in the six greenhouse gases. The EU target was 8 per cent. Japan's was 6. The overall target for the industrialized countries, including permitted increases for Australia, Iceland and Norway, amounted to 5.2 per cent.

This was not bad at all. It would surely send a signal to the energy markets, notwithstanding the potential loopholes inherent in trading and sinks.

And nobody could oppose this: it was the heart of the compromise that had made a deal possible.

'The overall target of 5.2 per cent is 30 per cent below business as usual,' Estrada said. It would be 10 per cent below expected emissions in the year 2000. 'This we can celebrate.' It *will*, he emphasized, have an impact on atmospheric greenhouse-gas concentrations.

Then, with the gavel blow adopting Annex B, Estrada made history. 'I will now forward this protocol for adoption by the plenary,' he said, beaming.

The applause was long and loud. On the platform, staid UN officials embraced. On the negotiating floor, I later learned, seasoned diplomats were weeping.

Don Pearlman, saying not a word to his colleagues, put his arms on his knees, extended his fingers, and rested his head on them, staring into nothing for several minutes. The rest of the carbon club sat, wordless and grim-faced, staring at the TV screen.

Epilogue

A month before the end of the decade, century and millennium, the world learned that the ice cap on the Arctic Ocean is melting fast. American researchers released ice-thickness data, gathered by nuclear submarines, that the Pentagon had been sitting on for years. It showed that over the last forty years the ice depth in all regions of the Arctic Ocean has declined by some 40 per cent.

This confirmation of fears I had harboured for more than a decade soured any sense of history or occasion that I might have felt about the end of the millennium. All those celebrations of 1,000 years of human history; what could they really amount to when the existence of civilization beyond even the next century is so open to doubt? I thought about what the freshwater released from the Arctic ice would end up doing to ocean circulation. I imagined the warming Arctic waters seeping their warmth into seafloor methane hydrates. The regular drip of other climatic worry stories fed my feeling of disenfranchisement. Before November was through, I read that British scientists had reported decreasing salinity in deep currents flowing south between Shetland and the Faroes. And in the first week of December I learned that cod had virtually disappeared in the North Sea not just because of overfishing but, so UK government scientists now thought, because the waters have warmed fully 4°C in the last six years. The warming water has been dividing into layers, meaning that there are fewer phytoplankton for the fish to feed on.

The Arctic news made me not just worried, but furious. I recalled how almost exactly ten years before, I had spent a morning leading a global warming seminar for a roomful of senior British military officers at Cambridge University. Among them had been two ballistic missile submarine commanders. That day, I had encountered a willingness to accept my arguments that was utterly unfamiliar at the time. The IPCC had not even finished its first scientific report, the one that was to have such an impact in the spring of 1990. The previous day, by contrast, I had given a similar seminar to a hall full of Ford motor workers at the car company's plant in Dagenham. I had met a wall of denial and hostility, typical of the industry response at the time. Why was it, I remember thinking, that military people were so much more able to accept the global-warming threat, and so much keener to talk about solutions? Could it be that they knew something I didn't?

Ten years on, I now know that they did. That submarine data on Arctic ice thickness should have been released to the imperilled world the moment the first IPCC report came out in 1990. But it took the military ten years to do the obvious.

Just as it took Ford Motor Company ten years to break out of its global-warming denial. On 6 December, the company became the first US-based company to quit the Global Climate Coalition. 'It has become to us an impediment to move forward credibly on environmental issues,' said Chairman William Clay Ford, Jr, by way of explanation. 'We're quite disappointed,' said my old enemy in the Coalition, Constance Holmes. I wondered if she and the other hatchetmen and women for Ford, Exxon and the rest were having their own consciences stirred by the flood of evidence now emerging.

1998 broke all records for global average annual surface temperature by a depressingly wide margin. The first nine months of 1998 one by one broke monthly records for the global hottest-ever. 1999 turned out to be the fifth hottest year. The ten hottest years in 138 years of records have all now been since 1983. As we approach the end of the century, the global average temperture is a full 0.7°C warmer than it was a century ago.

During the course of 1998 and 1999, climate scientists became more and more alarmed, and increasingly willing to voice their fears. In 1998, a new model by the UK's Meteorological Office suggested that in a world without deep emissions reductions, warming will kill many tropical forests in the second half of the twenty-first century, returning a vast quantity of carbon to the atmosphere. This would run the risk of tipping the world into runaway global warming. A meeting of worried scientists convened in Paris to consider to what extent nature has planted such Doomsday devices around the planet, awaiting the global-warming trigger. Top of the worry list was methane hydrates. Finally, it seems, scientists are losing their reluctance to articulate the worst-case analysis. As Stanford University climate guru Professor Steven Schneider told *New Scientist* magazine: 'We used to discuss these scenarios privately. Now we are being more open.'

The IPCC has announced that it will make sudden and chaotic climate change a central feature of its third report, due out in 2001. The second report, in 1995, had precisely two paragraphs on the subject. By November 1999, early drafts of chapters from the third report were already in circulation, and being leaked to the press. IPCC scientists seem unwilling to wait until 2001, and it is little wonder. Just before the last Christmas of the century, the chief meteorologists of the US and UK took the unusual step of writing to national newspapers to say that the recent work of their agencies 'confirms that our climate is now changing rapidly'. The situation, they warn, is critical.

It is becoming so obvious now. The ice shelves on the Antarctic Peninsula are in full retreat, having lost 3,000 square kilometres of ice in 1998 alone. The Greenland ice cap is also melting fast. Laser altimetry measurements published in March 1999 show rapid thinning in the eastern portion of the sheet, particularly in coastal regions, at rates exceeding a metre per year in some regions. An early draft of the third IPCC report concludes that a 3°C warming over Greenland would make melting of the ice cap irreversible, meaning a seven-metre rise in global sea level over the next millennium. Seven metres. Good-bye every hydrocarbon-century coastal city.

Coral bleaching has become endemic in many reef areas. A survey of reefs in the Seychelles in January 1999 showed that 80 per cent of the coral is dead, with over 95 per cent in some areas. The reefs on the paradise island are on the verge of extinction. Reports from other areas, if less extreme, are uniformly worrying.

The hottest year ever coincided with the costliest year ever for insured losses from weather-related catastrophes. The catalogue of storms, floods, droughts and fires around the world in 1998 exceeded all the weather-related losses of the 1980s. Four more events joined the list of billion-dollar catastrophes: three storms in America, and terrible summer floods along China's Yangtze River. One of the American storms was September's $3.3 billion Hurricane Mitch, another dodged bullet for Florida, in that it caught only the Florida Keys. Another hurricane, October's Mitch, killed over 10,000 in Central America and devastated the entire economy of Honduras.

1999 was little better. The insurance industry dodged yet another bullet in the USA. Hurricane Floyd looked for a while like being a monster. An industry report earlier in the year had concluded that a single $100 billion hurricanee loss was now a possibility. Although weakening by the time it hit shore, the storm ravaged 16 states over a four-day period in mid-September. But the insurance bill panned out at less than $2 billion. No cities had suffered a direct hit.

The worst storm in two centuries hit France on Boxing Day 1999. More than 100 million trees fell. Photographs resembled scenes from the First World War. In the grounds of Versailles, as the manager of the monument described it, 200 years of history disappeared in an hour. Thousands of troops were mobilized. 400,000 homes were still without electricity a week later. Floodwater led to a serious incident at a nuclear power plant near Bordeaux.

Still, amazingly, at the time of writing the insurance companies show little sign of being willing to mount a concerted response to the global-warming threat.

Six weeks into the new century, I addressed a meeting of British pension fund managers who had assembled to consider the environment and investment. The issue was becoming a hot topic for them. Their interest was awakened primarily because investments in environmentally or socially screened accounts had more than tripled between 1995 and 1997, amounting to over $500 billion in 1997, and the trend was continuing. My presentation was predictable. They had, I said, another very strong reason to look at the environmental implications of investing a river of capital in traditional energy companies, as they routinely did. The next speaker was Dr Julian Salt of the insurance industry's own loss think-tank, the Loss Prevention Council. He agreed with me. The whole *modus operandi* of the industry, he said, was inherently flawed. The insurance industry took $2 trillion in premiums within a $20 trillion world economy. Much of this was invested in the largest companies in the world, and many of these were the carbon-fuel-based companies that threatened the insurance industry with bankruptcy, the banking industry with huge losses of building assets in which capital was tied, and the capital markets with climate-catastrophe-related meltdown. There was only one solution: deep cuts in greenhouse gas emissions – decarbonization, in other words. The investment implications were clear, Dr Salt said. Solar energy had a huge future, hand in glove with the digital revolution. He had a message for investors, and it was simple. He put up an overhead with four words: 'Buy silicon. Sell carbon.'

The investment managers in the room listening to this had almost $250 billion in assets under their collective management.

I was encouraged. Dr Salt was not alone in his thinking. In mid-January, an American voice on the line turned out to be my first formal request for an opinion about global warming from a mainstream investment analyst advising institutional shareholders. He wanted to know if I thought corporate liability with regard to climate change would become an issue. That was for him to decide, I responded. But he should be aware that the top twenty privately owned carbon-fuel producers distributed almost a quarter of all the fossil fuels burned. The top 122 carbon producers produced 80 per cent of the carbon going into the atmosphere from fossil fuels. The global-warming writing was becoming ever clearer on the wall, and who knew when the world might start looking back in anger. Would I have billions locked up in strand-worthy so-called assets like oil refineries or coal mines? No way.

I feel sure the financial dam could yet burst, in favour of sanity and the solar revolution.

Could the climate negotiations help this process, after the qualified success of Kyoto? The 1998 climate summit was in Buenos Aires. I was in America, taking four Swiss bankers on a tour of solar investment prospects. I watched the news wires from Buenos Aires with trepidation. Newspaper reports told of the usual fraying of tempers as the OPEC nations stalled and filibustered. Don Pearlman's orchestration of these tactics, widely known by now, was described openly in the press. As so often before, the carbon club's wrecking campaign combined with general US-led governmental foot-dragging to stall the climate negotiations. As the summit neared its conclusion, governments could still not agree even the most basic rules to enact the emissions reductions agreed in Kyoto.

On the last night, environment ministers saved a little face by listing over a hundred separate decisions that governments would have to make within the next two years if the Kyoto Protocol was to work. They also agreed that the Clean Development Mechanism should go ahead as soon as possible. This meant that companies would be able to make early investments in renewable energy projects in the developing countries as a contribution to emissions reductions at home, confident of credit when the full Kyoto rules were agreed. Although President Clinton signed the Kyoto Protocol during the week of the Buenos Aires talks, there was still little sign that the present US Senate would ratify it.

By March 1999, one year after the Kyoto Protocol opened for signature, 84 countries had signed it. But only seven had taken the far more important step of ratifying it. Fifty-five will have to do so before it comes into force. Environmentalists want this to take place no later than the Rio-plus-10 summit in 2002: ten years after the Convention on Climate Change was adopted at the Rio Earth Summit in 1992.

The 1999 climate summit was in Bonn, during late September and early October. I couldn't stay away, even though Solar Century's immediate business needs were pressing at the time. I needed to see firsthand what was happening.

I arrived in the vast Hotel Maritim, venue for the talks, to find that the so-called flexible mechanisms of the Kyoto Protocol – emissions trading, joint implementation and the Clean Development Mechanism – were bogging the negotiations down in potentially irresolvable detail. Delegates thronged in the corridors in the usual thousands. As they went to and fro between the many increasingly complex side sessions into which the talks had split, they walked past a long line of exhibits staged by a host of interest groups. In a prominent bottleneck in the corridor, the UK Met Office had mounted a video display of its latest climate model, the result of five years' effort to factor in carbon loss in warming forests and soils. A video on repeat cycle showed temperatures rising by as much as 8°C over landmasses by the end of the twenty-first century: an artful, pulsing display that showed a progressively reddening world map as global temperatures rose to unsurvivable levels. While this official horror story played on, hundreds of diplomats scurried past it in longwinded pursuit of issues amounting to fractions of percentage points on the already far too low Kyoto target of only 8 per cent of total global emissions. The disconnect was plain. The negotiations, I saw, were freezing up in a culture of complexity imbued essentially by the insistence of the United States of America on accountancy gambits to escape its obligations to cut domestic emissions.

But just as governments were stalled, the business attendees were making headway. I saw that since Kyoto, a wholly new and positive atmosphere had built up at the negotiations. I had heard that this was the case in Buenos Aires a year before. Now I saw it for myself. The majority of business attendees were now pushing for rules to be laid down for emissions-reductions. The atmosphere was more of business opportunity than legislative threat. The message coming over from the packed industry side-meetings was 'how do we do this? Show us how the Clean Development Mechanism and joint implementation will work.' Companies now seemed to be accepting the inevitable, and looking for competitive advantage. This cast the governmental logjam in a less gloomy light. Many people I talked to felt that if the governments could just keep their show on the road – keep the talks from falling apart – then industry momentum might yet swing the world into an emissions-reduction culture. This in turn might create the political space for governments to find a legislative way forward at the October 2000 climate summit in the Hague.

The carbon club's lobbyists attended few of the industry meetings. These days, they were in the clear minority. They ploughed the usual furrows, whispering their stalling conspiracies with the Saudis and Kuwaitis. But they seemed different now. Don Pearlman, who had for so long radiated a dark power in the corridors of the climate talks, now looked to me like a tatty and harassed old man. The vast majority of attendees simply ignored him.

And so, as we enter the new millennium, the new business agenda gathers pace. BP and Shell, continuing to compete with each other for the mantle of most responsible oil company, both committed in 1998 to cut their greenhouse-gas emissions by 10 per cent from 1990 levels – a target higher than that of most governments. In America, a new effort to create a rallying point for relatively progressive corporations, the Pew Center on Climate Change, is stimulating spectacular defections. Among the companies to have agreed concrete steps to cut emissions, under the Pew umbrella, are American Electric Power, Boeing, Enron, Lockheed Martin, 3M and United Technologies. As the Chief Executive of American Electric Power put it, it is no longer possible to say there is not a problem.

Others in the oil industry are beginning to see the writing on the wall. Most spectacularly, ARCO's Chief Executive Mike Bowlin gave a speech to an oil industry audience in January 1999 in which he said that his industry was now entering 'the last days of the age of oil.' Oil and gas companies face a critical choice, he said. 'Embrace the future and recognize the growing demand for a wide array of fuels; or ignore reality and slowly but surely be left behind.' Among this wide range of fuels, renewable energy of all sorts would be vital.

No less an organ than the *Wall Street Journal* converted to this view in September 1999. Having for years run editorials lambasting global warming as a scare-story ploy of environmentalists opposed to business, the *Journal* ran a special issue just before the Bonn climate summit entitled 'The Race to Profit from Global Warming'. The opening sentence read like a death knell for the carbon club. 'In major corners of climate America, it's suddenly become cool to fight global warming.' Sure enough, when the hundreds of chief executives at the World Economic Forum in Davos in February 2000 voted on the number one issue of concern to business in the future, climate change came top of the list.

In the face of this rising tide, the Global Climate Coalition's planet-wreckers guild continues to splinter. Daimler–Chrysler quit the GCC in the first week of January 2000. To add insult to injury, business wires began reporting an unprecedented surge in the value of renewable energy stocks around the same time.

These themes are increasingly being picked up by the media. On the very last day of the twentieth century I found myself live on BBC Radio's morning *Today* programme. In a strange microcosm of my previous decade, I was debating with an obscure apologist professor whether or not burning oil was even a problem. But sitting next to me in the studio, listening, was the Met Office's famous weather forecaster Michael Fish. As soon as the short debate was over, he read the weather forecast. But not before injecting a stiff comment to the effect that the recent hot years were a clear indication of global warming to come. Three days into the new century, I found myself live on CNN. For years I had been used to the American network treating the climate-change issue in an almost tongue-in-cheek way, allocating only a few minutes, and almost always pitting a contrarian against me. Now I found myself part of a one-hour special on global warming that took as its starting point the fact that the planet was in crisis over global warming, and what the hell were we going to do about it. I had a full ten minutes to spout about solar energy's role in the solution.

But a year on from Kyoto, the bottom-line question poses itself ever more starkly. Will – can – the engine of change move fast enough?

These days, I read about the march of events in the global-warming story in spare moments as I go about the business of trying to transform myself into a solar-energy entrepreneur. I try not to let the frustration I feel stand in the way of giving Solar Century's effort to accelerate the solar revolution my best shot. But there is no denying that my worries about just how dangerous global warming is, and the delay caused by the pulled punches of the climate scientists over the years, knock an edge off the adventure I am now embarked on in the energy markets.

But hope lives on. It lives in the seminal transformations that the increasingly inescapable reality of global warming is forcing in the more responsible corporations. It lives in the growing evidence of popular concern and desire for meaningful actions. During 1998, a survey showed that a clear majority of Texans now accept that global warming is a threat. Most of them say they would be prepared to pay more for renewable energy than fossil fuels in order to do something about it. I write this not to snipe at Texas but simply because such opinions would not have seemed possible there at the time of the Kyoto climate summit. It shows how experience counts. April through June was the hottest such period in over a century in Texas, with prolonged insufferable heat in dozens of cities right across south America.

In the year since Kyoto, I have had the roof of my own home turned into a solar power station. I am the proud owner of the UK's first solar PV rooftile home, and since I connected to the national grid in March 1999, I have generated more electricity than I have used in the house. My stunning-looking roof will keep more than thirty tonnes of greenhouse gas emissions out of the atmosphere during its lifetime. My company Solar Century is trying to sell other such roofs – notwithstanding the current price trap the technology languishes in – endeavouring to help accelerate the solar revolution. We may succeed, we may fail. But the fate of one small solar company is not the main issue. The fact is that there will be an inescapable explosive take-off point for solar PV sometime soon. As increasingly carbon-savvy investors capitalize more solar companies for expansion, so the price will come down. It is happening already. The global solar PV market grew by over 30 per cent in 1999. The solar revolution, like the enormity of the global-warming threat itself, is no longer in doubt.

As we approach the end of the hydrocarbon century, the oil companies shuffle for position, patently uncertain of the way forward as their world changes around them as never before. The low oil prices of 1998 drove them to start merging, so that the end of the century begins to look increasingly like the beginning: a global market dominated by three huge companies. Then it was Standard Oil, BP's forerunner, Anglo-Persian, and Shell. Now it looks like being Exxon–Mobil, BP–Amoco–Arco and Shell. Two of the three groupings have chosen to make encouraging noises about solar power. But they have gone little further than sticking solar panels on petrol stations so far. They deny for the moment that it is feasible to build a PV manufacturing plant big enough to make solar electricity as cheap as coal power. They are wrong. They could build such a plant for not much more than $100 million: less than the cost of a single leg of a solitary oil rig. This last stand won't endure. BP and Shell have created too much momentum – too much legitimacy for the technology – by their actions in refusing to emulate the tobacco industry when it comes to their core business and its role in global warming. Meanwhile, the oil price is shooting up again. As I write these words it passed the $30 per barrel mark.

And what of the footsoldiers of the carbon club? They will continue, no doubt, to try delaying the inevitable. Exxon, Mobil, Texaco and the other residually unrepentant thugs of the corporate world look like continuing to sign the cheques that bankroll the carbon club's crimes against humanity, along with their kindred spirits in the auto, coal and utility industries. They may well enjoy minor victories along the way. But they have already lost the pivotal battle in the carbon war. The solar revolution is coming. It is now inevitable.

The only question left unanswered is, will it come in time?

Jeremy Leggett
February 2000
Richmond, UK

Index